T0192574

Mechano-Electric Correlations in the Human Physiological System

Biomedical and Robotics Healthcare

Series Editor:

Utku Kose, Jude Hemanth, and Omer Deperlioglu

Artificial Intelligence for the Internet of Health Things

Deepak Gupta, Eswaran Perumal, and K. Shankar

Biomedical Signal and Image Examination with
Entropy-Based Techniques

V. Rajinikanth, K. Kamalanand, C. Emmanuel, and B. Thayumanavan

Mechano-Electric Correlations in the Human Physiological System

Dr. A. Bakiya, Dr. K. Kamalanand, and Dr. R. L. J. De Britto

For more information about this series, please visit:
https://www.routledge.com/Biomedical-and-Robotics-Healthcare/book-series/
BRHC

Mechano-Electric Correlations in the Human Physiological System

A. Bakiya, K. Kamalanand, and R. L. J. De Britto

CRC Press
Taylor & Francis Group
Boca Raton London New York

CRC Press is an imprint of the
Taylor & Francis Group, an **informa** business

MATLAB® is a trademark of The MathWorks, Inc. and is used with permission. The MathWorks does not warrant the accuracy of the text or exercises in this book. This book's use or discussion of MATLAB® software or related products does not constitute endorsement or sponsorship by The MathWorks of a particular pedagogical approach or particular use of the MATLAB® software.

First edition published 2021
by CRC Press
6000 Broken Sound Parkway NW, Suite 300, Boca Raton, FL 33487-2742

and by CRC Press
2 Park Square, Milton Park, Abingdon, Oxon, OX14 4RN

© 2021 Taylor & Francis Group, LLC

CRC Press is an imprint of Taylor & Francis Group, LLC

The right of Dr. A. Bakiya, Dr. K. Kamalanand, and Dr. R. L. J. De Britto to be identified as authors of this work has been asserted by him/her/them in accordance with sections 77 and 78 of the Copyright, Designs and Patents Act 1988.

Library of Congress Cataloging-in-Publication Data
Names: Bakiya, A., author. | Kamalanand, K., 1988– author. | DeBritto, R., author.
Title: Mechano-electric correlations in the human physiological system /
A. Bakiya, K. Kamalanand, R. DeBritto.
Description: First edition. | Boca Raton, FL : CRC Press, 2021. | Includes
bibliographical references and index. |
Summary: "The aim of this book is to present the mechanical and electrical properties of human soft tissues and the mathematical models related to the evaluation of these properties in time, as well as their biomedical applications. The book also provides an overview of the bioelectric signals of soft tissues from various parts of the human body"—Provided by publisher.
Identifiers: LCCN 2020049772 (print) | LCCN 2020049773 (ebook) |
ISBN 9780367622626 (hardback) | ISBN 9781003109181 (ebook)
Subjects: LCSH: Tissues—Mechanical properties—Mathematical models. |
Electrophysiology—Mathematical models. | Relaxation phenomena. |
Electromyography. | Human physiology.
Classification: LCC QP88 .B28 2021 (print) | LCC QP88 (ebook) | DDC 612—dc23
LC record available at https://lccn.loc.gov/2020049772
LC ebook record available at https://lccn.loc.gov/2020049773

ISBN: 978-0-367-62262-6 (hbk)
ISBN: 978-1-003-10918-1 (ebk)

Typeset in Times
by codeMantra

Contents

Preface

The human physiological system is highly complex consisting of various coordinated subsystems with several intricate interrelationships. The analysis of such interrelations at the cell, tissue and organ level is decidedly important to understand the functioning of the physiological system and, in turn, to design medical devices for the diagnosis and treatment of various diseases. The measurement, modeling and analysis of the dielectric and mechanical properties of the soft tissues provide an insight into the interactions of the physiological systems with external electric fields, as well as into the effect of various external forces on the human body. Further, the measurement of such properties provides new methods and markers for the diagnosis of various soft tissue pathologies. Further, there is a close correlation between the electrical and mechanical properties of biological tissues. The estimation of such correlations is expected to provide leads to further understand the disease mechanisms, its severity, prognosis and the effect of treatment.

The branch of electrophysiology describes the recording of electrical signals generated by the physiological systems, the associated instrumentation, the mathematics and the methods of analysis of such signals for supporting the clinicians to clinch the diagnosis of the diseases and to prescrble an appropriate treatment plan. The biosignals recorded from individual provide useful information on the mechanical activity of the organ systems generating the signals. However, the analysis of such signals is a complex task and requires computer-aided systems for automated diagnosis and mass screening of various diseases.

This book entitled *Mechano-Electric Correlations in the Human Physiological System* presents the basic mechanical and electrical properties of soft tissues, the mathematical models related to the evaluation of these properties in time and frequency domains along with their interrelationships. Further, this book describes the electrophysiological aspects of the physiological system with a specific focus on electromyograms (EMGs) along with case studies involving the effects of fibrotic lymphoedema on the measured EMG signals.

The book is organized into five chapters. The first chapter presents the characteristics of soft tissues, the viscoelastic and dielectric properties of soft tissues along with the mathematical models for the analysis of such properties. This chapter also provides simple MATLAB® programs for the simulation analysis of the viscoelastic and dielectric characteristics of the soft tissues, both in time and frequency domains.

Chapter 2 presents the interrelationship analysis on the viscoelastic and dielectric properties of human soft tissues with case studies involving liver and brain tissues. Further, it presents the methods and results involving the prediction of the mechanical and electrical properties of soft tissues using machine learning approaches such as artificial neural networks and support vector machines.

Chapter 3 presents an introduction to electrophysiology, action potentials and the Hodgkin–Huxley model of action potentials. Further, the characteristics and classification of biosignals, the problems associated with recording biomedical signals, the physiological transducers, electrode types and characteristics, the model for analyzing the skin-electrode interaction, the typical physiological biosignals such as the electrocardiograms and the EMGs along with the instrumentation systems for acquisition of the biosignals are described in detail.

Chapter 4 presents an introduction to the neuromuscular system and EMG, the effects of various types of EMG electrodes such as surface electrodes and needle electrodes on the acquired EMG signals, the placement of electrodes for EMG recording, and the computational analysis on normal, myopathic and neuropathic EMG signals along with appropriate results.

The final chapter describes the characteristics of EMGs acquired from muscles in a non-myopathic condition lymphatic filariais most prevalent in developing countries. This chapter presents the different stages of lymphoedema in lymphatic filariasis patients and the fibrotic changes occurring in the affected limb. Further, it provides an insight into the preliminary investigation on EMG signals acquired from normal leg and fibrotic lymphedema leg using time domain and frequency domain analyses.

This book is designed as a reference material for researchers, students and academicians involved in electromechanical and electrophysiological studies on soft tissues. This book also aims to describe various mathematical techniques for the analysis of the material properties of soft tissues with relevant examples. We hope that the readers find this book interesting and informative.

A. Bakiya
K. Kamalanand
R. L. J. De Britto

MATLAB® is a registered trademark of The MathWorks, Inc. For product information, please contact:

The MathWorks, Inc.
3 Apple Hill Drive
Natick, MA 01760-2098 USA
Tel: 508-647-7000
Fax: 508-647-7001
E-mail: info@mathworks.com
Web: www.mathworks.com

Acknowledgement

The authors would like to thank Dr M. K. Surappa, the Vice-Chancellor of Anna University, Chennai, Dr L. Karunamoorthy, the Registrar of Anna University, Chennai and Dr Ashwani Kumar, Director, Vector Control Research Centre (VCRC), Indian Council for Medical Research (ICMR), Puducherry, for their constant support and encouragement. The authors express their gratitude to the Editor Dr Marc Gutierrez at CRC Press for his continuous support right from the beginning to the publishing stage. The authors would like to thank the Series Editor, Dr Utku Kose, for accepting the book in this series. The authors thank Nick Mould, Editorial Assistant, CRC Press for his valuable inputs at the stage of final publication. It was really a pleasant experience working with the team throughout this project. The authors acknowledge their family members for their cooperation to the untoward situation due to COVID-19 pandemic and, in particular, the widespread pandemic in Chennai, a metropolitan city in India.

A. Bakiya
K. Kamalanand
R. L. J. De Britto

Authors

Dr. A. Bakiya received her M.E. degree in VLSI Design from Anna University and completed her Ph.D. in the field of biosignals and fractional calculus at MIT Campus, Anna University. She has 5 years of teaching experience and is a recipient of the "Excellence in Teaching Award" from SKP Engineering College, Tiruvannamalai. At present, she is a co-investigator in three projects in the field of Electrophysiology and Infectious Diseases. She has served as an ad-hoc reviewer for various edited books and journals. She has the credit of a co-inventor for an invention and three designs in the field of medical instruments. She has co-authored several book chapters, research articles in international journals and conference proceedings.

Dr. K. Kamalanand completed his Ph.D. at MIT Campus, Anna University in the field of Intelligent modeling and analysis of HIV/AIDS infection. At present, he is an Assistant Professor at the Department of Instrumentation Engineering, Madras Institute of Technology Campus, Anna University, Chennai, India. He has published four books, six chapters in edited books, 35 research articles in SCI/Scopus journals, 31 articles in international conference proceedings and seven research articles in national conference proceedings. He has three awarded patents and four published patents in the field of biomedical engineering. He has served as a guest editor for the *European Journal for Biomedical Informatics* (Official journal of the European Federation for Medical Informatics), *Current Bioinformatics* and *Current Signal Transduction Therapy.* He is a member of the Council of Asian Science Editors. He is a Fellow of the International Society of Biotechnology.

 **Dr. R. L. J. De Britto** obtained his M.D. post-graduate degree from Government Stanley Medical College and has 35 years of research experience in the fields of clinical trials, vaccine trials and infectious diseases. His major focus of research interest is infectious diseases, in particular, leprosy, tuberculosis and filariasis. He was instrumental in establishing two institutes, National Institute of Epidemiology, Chennai and National Institute of Traditional Medicines, Belgaum. He was trained in HIV/AIDS Epidemiology in Developing Countries under Fogarty Fellowship in the School of Public Health, the University of California, Los Angeles (UCLA), United States and in International Communicable Diseases, under WHO – Fellowship in Ministry of Public Health (MoPH), Thailand. At present, he is Scientist F (Medical) in the Indian Council of Medical Research (ICMR), the apex body for the biomedical research in India. He has evaluated the National Programme of Elimination of Lymphatic Filariasis in India. Currently, he is involved in the assessment of fibrosis in lymphedema of the extremities and has developed a therapeutic device in filarial lymphedema and its associated medical conditions. He is a member of the International Infectious Diseases and Indian Society of Leprosy and has received the Ackworth award for the best-published paper conferred by the Indian Association of Leprologists. He has published more than 40 research papers in national and international journals.

1 Mathematical Modeling and Analysis of Soft Tissue Viscoelasticity and Dielectric Relaxation

1.1 MATHEMATICAL MODELING IN MEDICINE AND BIOLOGY

Mathematics is considered as the backbone of engineering and technology and provides a strong framework for the design and development of real-time devices and instruments. In fact, all real-world applications are governed by a set of mathematical equations which represent the principles and operations of the equipment. Mathematical modeling is the art of representing physical systems and processes using functional expressions operating on a set of mathematical operations (Kamalanand & Jawahar, 2018). These models are highly useful for the design, prototyping, development and maintenance of the systems under consideration (Krishnamurthy, 2016; Krishnamurthy & Jacob, 2014).

In the fields of medicine and biology, computational techniques are of great help for the analysis of biological systems (Edelstein-Keshet, 1988; Chaplain, 2011; Bailey, 1975; Rideout, 1991). The applications of mathematical modeling in the field of biomedical engineering and technology include:

1. *Diagnostics*: Mathematical models are developed for identification of a disease or a disorder of physiological systems (Barnabas et al., 2006; Kamalanand and Ramakrishnan, 2015; Schwarzer et al., 2000; Khoo, 2018). Because of the spread of diseases such as H1N1 and Ebola in epidemic proportions, it is highly useful to develop mathematical frameworks for the fast and efficient diagnosis of these diseases using available data. Such systems are termed as "mass screening systems" and are greatly helpful in resource-limited countries as well as in countries with large populations, where individual attention during an epidemic is challenging.

2. *Therapeutics*: In certain cases, it is necessary for a proper scheme to administer therapy efficiently. For example, in an HIV/AIDS-infected patient, antiretroviral drugs are used as a drug for providing therapy. In addition, a combination of antiretroviral drugs, identified as highly active antiretroviral therapy is often utilized. Hence, there is a need to optimize the amount of drugs used to efficiently administer therapy (Kamalanand and Jawahar, 2016). If a large amount of drugs is used, the side effects may be severe, and if too little of the drug is used, it may not be effective in reducing

1

the HIV viral load (number of virus particles in 1 μL of blood) to the desired value. Hence, mathematical optimization techniques play a vital role in the development of effective strategies for therapy planning in certain diseases (Granich et al., 2009; Tanaka et al., 2010; Raphael, 1992; De Pillis and Radunskaya, 2001).

3. *Prosthetics*: Prosthetics involves the design and development of artificial devices for the replacement of lost body parts. These devices are functionally similar to that of the original body part. For example, artificial limbs developed to replace amputated or lost arms or legs are capable of performing the original operations. This is possible due to the combination of electrophysiological measurements, robotics and fast-processing units coordinated through strong mathematical and computational pillars (Johnston et al., 1979; Watton et al., 2007; Stansfield et al., 2003; Pitkin et al., 2009).

Other computational applications in the field of medicine involve the assessment of severity of certain diseases, epidemiological analysis (Ambikapathy and Krishnamurthy, 2020; Krishnamurthy et al., 2020), biomechanical analysis, medical image and signal analysis, prediction and classification of medical data, etc. (Krishnamurthy and Ponnuswamy, 2018; Kamalanand et al., 2011; Kumar et al., 2016; Kamalanand and Jawahar, 2015a,b; Manickavasagam et al., 2014). In recent years, finite element methods have been employed for analyzing the effect of mechanical stresses on soft and hard tissues as well as for the design of diagnostic equipment for the analysis of various physiological parameters (Arunachalam et al., 2014; Kamalanand and Srinivasan, 2011a,b). Mathematical methods of analysis employ a computer-aided framework for finding the solutions to complex biomechanical problems such as the assessment of viscoelastic characteristics of human soft tissues.

The process of mathematical modeling in medicine and biology involves the development of suitable mathematical descriptions or functional representations which include the real-world parameters, and the output of these models need to closely match the experimental results. There are two methods of mathematical modeling. In the first method, the model is developed using a thorough knowledge of the physical principles governing the biological system. Such models are known as the first principle models. For example, the Maxwell model is a first-principle model of the viscoelasticity of biological soft tissues. The process of development of such models is described in Figure 1.1.

In the second method, the results of the experiments conducted on the biological systems are utilized to develop mathematical equations to describe the system. Curve fitting is an example of this method. In this technique, the results of the real-world experiments are utilized to estimate the parameters of a known mathematical function for describing the system or process. This technique utilizes an optimization algorithm for obtaining the model parameters. Figure 1.2 shows the process of developing mathematical models using the second method.

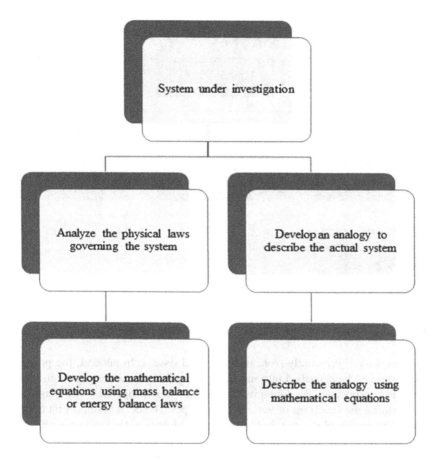

FIGURE 1.1 Mathematical modeling using first principles approach.

1.2 SOFT TISSUE CHARACTERISTICS

The tissues in the human body can be classified into soft and hard tissues. The soft tissues play a significant role in the proper structural and functional characteristics of the physiological system. The various soft tissues in the human physiological system include the tendons, muscles, ligaments, fascia, skin, nerves, fibrous tissues, fat and blood vessels. Soft tissues is are classified into connective and non-connective tissue (Figure 1.3).

The functions of the soft tissue include providing connections, providing support and holding other organs in specific places in the human body. Thus, soft tissues play an important role in providing structure and symmetry to the human body.

On the contrary, hard tissues are calcified tissues in the human body, which are formed by the mineralization process. The various hard tissues in the body are bone, tooth enamel, dentin and cementum. The material properties and characteristics of

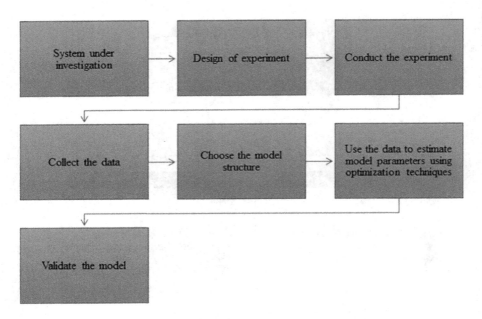

FIGURE 1.2   Mathematical modeling using optimization techniques.

soft tissues vary significantly from those of hard tissues. In general, the properties of a material can be classified into mechanical, optical, acoustic, electrical, thermal and magnetic properties, as shown in Figure 1.4. These properties play a significant role in defining the functions of various tissue types. Hence, it is important to have a thorough knowledge of the material properties of biological tissues to design suitable therapeutic and diagnostic equipment.

The various mechanical properties of biological soft tissues include density, Young's modulus or elastic modulus, viscous modulus or loss modulus, breaking stress, breaking strain, hardness, etc. These properties can be assessed using suitable mechanical tests and instruments such as the universal testing machine, the indentation device for measuring hardness and impact tester. The common electrical properties of the tissues are its resistivity, conductivity, permittivity and permeability. Similar to other material properties, these properties also vary significantly in cases of pathologies and diseases and are useful for assessing the quality of physiological systems. The thermal properties of soft tissues include thermal conductivity, thermal diffusivity, etc. and have a major influence on the functionality of the soft tissue. This book mainly deals with the mechanical, electrical and thermal along with the computation methods and models used to analyze the conditions of the tissue based on these properties.

1.3   PHYSICAL MODELS OF SOFT TISSUES

Tissue-mimicking materials are frequently utilized as phantoms or physical models to study, understand and simulate the properties of human and animal soft tissues (Madsen et al., 1982). In medical research, phantom materials are utilized

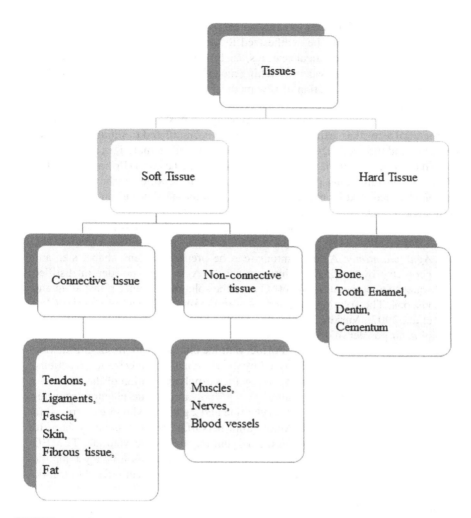

FIGURE 1.3 Classification of biological tissues.

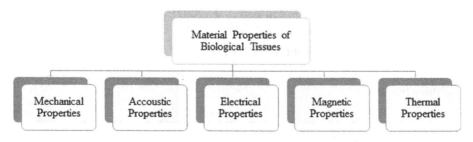

FIGURE 1.4 Material property classification of soft tissues.

as a substitute for soft tissues in cases where *in-vivo* tissues are unavailable. Tissue-mimicking phantoms can be synthesized to mimic the material properties as well as the geometry of anatomical features, such as blood vessels and internal organs (Surry et al., 2004). The tissue-mimicking materials are crucial in clinical studies for the development and validation of new medical devices as well as for the calibration of diagnostic and surgical instruments (Takegami et al., 2004; Erkamp et al., 2004). Various tissue-mimicking phantoms such as agar, epoxies, polymers, urethanes and other natural as well as artificial materials have been utilized to simulate the material properties and the geometry of soft tissues. Some patented materials, such as Zerdine which has properties similar to human liver tissue, exist as well (Rowan and Pedersen, 2006). Tissue-mimicking polyacrylamide gels were reported to be well suited for both bioelectrical and biomechanical analysis as these properties are tunable in the physiological range (Kao et al., 2008; Krishnamurthy et al., 2009). In this section, the preparation of two different tissue-mimicking phantom materials is discussed.

a. *Agar phantoms*: Agar phantoms can be prepared in any shape, size and geometry by heating a solution made of Agar Agar powder in distilled water to a temperature of 160°C until the solution becomes transparent and viscous. The solution is then cooled to obtain the phantom material (Zell et al., 2007). Various tissues can be modeled by varying the concentration of Agar powder in distilled water.

b. *Polyacrylamide phantoms*: Polyacrylamide phantoms are tissue-mimicking materials which are synthesized by the chemical polymerization reactions. These phantoms are synthesized by the co-polymerization of the monomer acrylamide and bis-acrylamide. A 40% polyacrylamide phantom requires a mixture of 38 g monomer acrylamide and 2 g bis-acrylamide dissolved in 100 mL deionized water. Ammonium persulfate is used as an initiator to initiate the reaction between the acrylamide and bis-acrylamide. TEMED (N, N, N', N'-tetramethylethylenediamine) is added as a catalyst for the polymerization reaction. Polyacrylamide gels can be synthesized to mimic the material properties such as the viscoelastic, dielectric and optical properties of several soft tissue types by varying the concentration of the monomer acrylamide (Kamalanand et al., 2010).

Further, phantom materials such as Silicone phantoms and polyvinyl alcohol gels can be synthesized in such a manner that their properties are similar to that of real soft tissues using procedures described in Zell et al. (2007).

1.4   SOFT TISSUE VISCOELASTICITY

In general, soft tissues are viscoelastic and anisotropic. Elasticity can be classified into three types:

i) *Elastic materials*: Elastic materials have a tendency to return to their original shape after applied forces are removed. Such materials are represented using a Hookean spring.

ii) *Viscoelastic materials*: Viscoelastic materials demonstrate both viscous as well as elastic characteristics when subjected to deformation due to stress. Such materials are represented using a combination of Hookean springs and Newtonian dashpots.

iii) *Hyperelastic materials*: These materials, unlike true elastic materials, do not have linear stress–strain relationships. Such materials are characterized by their non-linear stress–strain behavior.

In soft tissues, viscoelastic properties such as creep and hysteresis can be experimentally observed (Findley and Davis, 1989). Moreover, the stress–strain curve of soft tissues are non-linear in nature, as shown in Figure 1.5.

1.4.1 Viscoelastic Testing of Soft Tissues

The biological soft tissues are viscoelastic in nature and the following three different tests are in practice:

1. Creep test
2. Stress relaxing test and
3. Dynamic loading test

These tests are used to characterize the material. In the creep test, the strain $\varepsilon(t)$ resulting from the application of a steady stress σ_0 along only one axes is measured as a function of time, as shown in Figure 1.6.

In the stress relaxation test, the time-dependent stress resulting from a steady strain is measured, as shown in Figure 1.7.

The creep test and stress relaxation tests are highly useful to analyze the response of the material for longer times in minutes or hours; however, these tests are not suitable for short times or a few seconds. In dynamic tests, the strain resulting from a sinusoidal stress is monitored and recorded. When a viscoelastic soft tissue is subjected to a sinusoidal stress, the resulting strain will also be sinusoidal with the same angular frequency (ω) but with a different phase angle δ. The stress and strain function scan be given by:

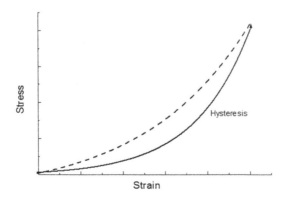

FIGURE 1.5 Typical stress–strain relationship in soft tissues.

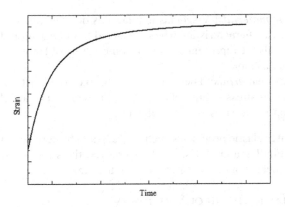

FIGURE 1.6   Results of the creep test.

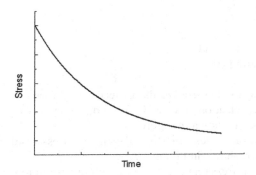

FIGURE 1.7   Results of the stress relaxation test.

$$\varepsilon = \varepsilon_0 \cos \omega t$$

$$\sigma = \sigma_0 \cos(\omega t + \delta)$$

Expressing it as a complex quantity, we get:

$$\sigma^* = \sigma_0' \cos(\omega t) + i\sigma_0'' \sin(\omega t)$$

where, $i^2 = -1$ and,

$$\tan \delta = \frac{\sigma_0''}{\sigma_0'}$$

$$\left|\sigma^*\right| = \sigma_0 = \sqrt{\sigma_0'^2 + \sigma_0''^2}$$

$$\sigma_0' = \sigma_0 \cos \delta$$

$$\sigma_0'' = \sigma_0 \sin \delta$$

and the storage modulus is defined as:

$$E' = \frac{\sigma_0'}{\varepsilon_0}$$

and the loss modulus is defined as:

$$E'' = \frac{\sigma_0''}{\varepsilon_0}$$

Dynamic analysis of soft tissues is performed using a dynamic mechanical analyzer (DMA). Using this instrument, two different types of tests can be performed. In the first test, the material can be subjected to sinusoidal stress and the resulting strain can be measured. The resulting strain, which is also sinusoidal, is phase-shifted by a factor. In a DMA analyzer, the sinusoidal stress is generated by a motor which generates a sinusoidal wave and transmits it to the sample. The resulting displacement of the sample is measured using a linear variable displacement transducer. Because a sinusoidal force is applied, the modulus can be expressed as a complex variable with an in-phase component known as the elastic or storage modulus and an out-of-phase component known as the loss modulus or the viscous modulus. In addition, the second type of test known as the temperature sweep test, where the storage and loss modulus of the sample is measured at a stress of low frequency at different values of temperatures, can also be performed. This test provides information on the effects of temperature variations on the elastic and viscous nature of soft tissues. The schematic of the DMA analyzer is shown in Figure 1.8.

1.5 MATHEMATICAL MODELING OF SOFT TISSUE VISCOELASTICITY

The results of a tensile test or compression test on a strip of soft tissue can be obtained through the simulation of a simple model of the tissue identified as the Maxwell model, consisting of a spring and a damper (Figure 1.9). The spring models the elastic nature of the soft tissue and the damper models the stiffness of the spring, and damping coefficient of the damper is known. The variation of the length (x) of the tissue on the application of a tensile or compressive load can be determined by solving the differential equation of the system, as shown in Figure 1.9.

$$\eta \frac{dx}{dt} + \mu x = F \tag{1.1}$$

In this model, the stiffness (μ) of the spring is related to the storage modulus or Young's modulus of the tissue under investigation, and the damping coefficient (η) is a measure of the viscous modulus or loss modulus of the soft tissue considering its dimensions. Applying Laplace transform on Equation (1.1),

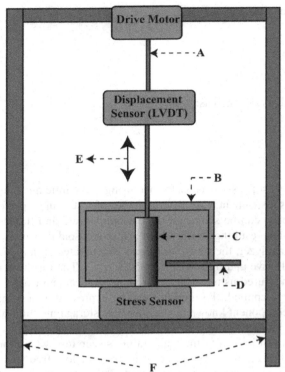

A - Drive Shaft; B - Furnace; C - Sample to be tested; D - Temperature Sensor
E - Movement Direction; F - Guidence System

FIGURE 1.8   Schematic of a dynamic mechanical analyzer.

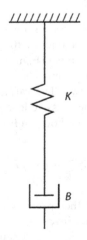

FIGURE 1.9   Maxwell model.

$$\eta s X(s) + \mu X(s) = F(s)$$

$$X(s)\big(\eta s + \mu\big) = F(s)$$

$$\frac{X(s)}{F(s)} = \frac{1}{\eta s + \mu} = \left(\frac{1/\mu}{(\eta/\mu)s + 1}\right)$$

This representation is known as the transfer function model and can be used to study the reaction of the system with various types of loads such as the Heaviside function, impulsive loads and dynamic or sinusoidal loads, as shown in Figure 1.10. The quantity $\tau = \dfrac{\eta}{\mu}$ is known as the time constant (τ) of the system and is measured in seconds.

Suppose, $F(s)$ is a simple Heaviside function whose Laplace transform is $\dfrac{1}{s}$, we get,

$$X(s) = \frac{1}{\big(\eta s + \mu\big)} \cdot \frac{1}{s}$$

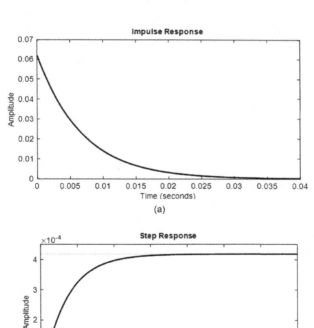

FIGURE 1.10 Unit impulse response (a) and unit step response (b) of the Maxwell model transfer function of brain tissue.

MATLAB® PROGRAM FOR OBTAINING THE IMPULSE AND STEP RESPONSE OF THE MAXWELL MODEL TRANSFER FUNCTION

```
mu=2.39e3;
eta=16.1;
n=[(1/mu)];
d=[(eta/mu} 1];
maxwell=tf(n, d);
figure(1)
impulse(maxwell);
figure(2)
step(maxwell);
```

The viscoelastic properties of biological soft tissues are crucial to physicians and surgeons as it provides valuable information on the physiological and pathological conditions of soft tissues. The mechanical properties of normal and abnormal soft tissues need to be investigated for various medical applications such as biomechanical simulation, development of diagnostic and therapeutic devices and planning surgery (Zhang et al., 2007). Various methods such as magnetic resonance elastography (MRE), ultrasound-based imaging methods, indentation devices and tactile sensors are available for the experimental investigation of the viscoelastic characteristics of biological soft tissues (Dhar and Zu, 2007). Often, diseases, such as various malignancies and liver cirrhosis, are diagnosed using viscoelastic parameters.

The viscoelastic properties of biological tissues are usually modeled as a combination of mechanical elements such as the spring and dashpot as it provides a simple and efficient representation (Zhang et al., 2007). The classical models which describe the linear viscoelastic behavior are the Maxwell model, Kelvin–Voigt model, and the standard linear solid model. These models are developed using the serial and parallel combination of springs and dashpots (Fung, 2013). Later, the fractional calculus was introduced into viscoelastic models. A modified Kelvin–Voigt model consisting of a spring in parallel to a dashpot with the stress in the dashpot equal to the fractional derivative of order "α" of the strain. This model is known as the Kelvin–Voigt fractional derivative model which incorporates the fractional derivative in the Kelvin–Voigt model (Caputo et al., 2011; Kiss et al., 2004).

In modeling soft tissue mechanics, in addition to the Maxwell model, four other models, namely, the Voigt model, the Zener model, the Jeffrey's model and the fractional Zener model are commonly used (Klatt et al., 2007). These models include a combination of one or more springs and dampers in series or parallel, as shown in Figure 1.11.

In general, the mechanical modulus of the soft tissue is expressed as a complex variable, given as a function of frequency (ω). That is, the storage modulus and the viscous modulus are both expressed as functions of frequency. The models given in Figure 1.11 are expressed using frequency-dependent functions in Table 1.1.

As mentioned, the mechanical modulus of the material is expressed as the complex variable $G_M(\omega) = G_1(\omega) + iG_2(\omega)$. The real part $G_1(\omega)$ of the complex modulus is the

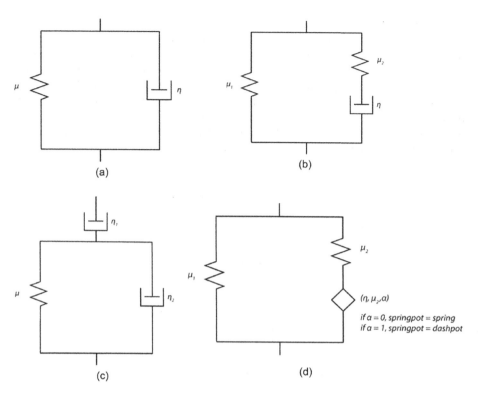

FIGURE 1.11 (a) The Voigt model, (b) The Zener model, (c) The Jeffrey's model and (d) the fractional Zener model.

TABLE 1.1
Viscoelastic Soft Tissue Models and Their Dynamic Modulus

Model Type	Complex Modulus Expression
Maxwell model	$G_M(\omega) = \dfrac{i\omega\eta\mu}{\mu + i\omega\eta}$
Voigt model	$G_M(\omega) = \mu + i\omega\eta$
Zener model	$G_M(\omega) = \dfrac{\mu_1\mu_2 + i\omega\eta(\mu_1 + \mu_2)}{\mu_2 + i\omega\eta}$
Jeffrey's model	$G_M(\omega) = -\omega\eta_1 \dfrac{\omega\eta_2 - i\mu}{\mu + i\omega(\eta_1 + \eta_2)}$
Fractional Zener model	$G_M(\omega) = \mu_1 + \dfrac{\mu_2\left(\dfrac{i\omega\eta}{\mu_2}\right)^\alpha}{1 + \left(\dfrac{i\omega\eta}{\mu_2}\right)^\alpha},\quad 0 \le \alpha \le 1$

elastic or storage modulus and is a function of frequency (ω). The imaginary part $G_2(\omega)$ of the complex modulus is the viscous or loss modulus and is also a function of frequency (ω). If the parameters of any of the viscoelastic models are known, then the model equations given in Table 1.1 can be simulated to obtain the values of storage and loss modulus at different frequencies. Figures 1.12 and 1.13 show the variations of the elastic and viscous modulus of different tissues such as the brain and liver tissues, as a function of frequency, respectively. Table 1.2 presents the values of parameters of the Zener model for normal and fibrotic liver tissues based on the experimental investigations conducted by Asbach et al. (2008).

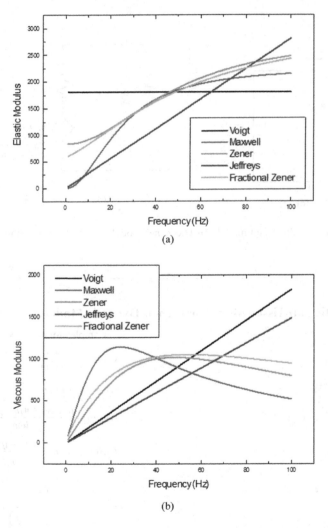

FIGURE 1.12   Variation of (a) elastic modulus and (b) viscous modulus shown as a function of frequency in brain tissue.

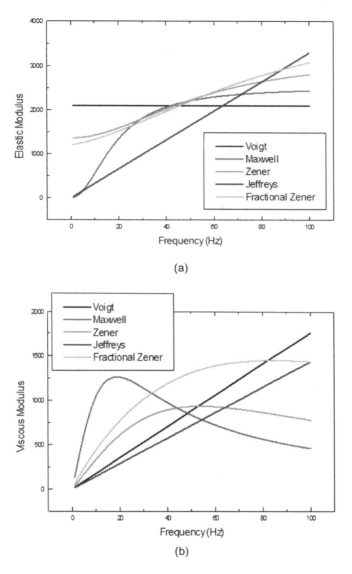

FIGURE 1.13 Variation of (a) elastic modulus and (b) viscous modulus shown as a function of frequency in liver tissue.

Both the liver and brain are challenging targets for rheological analysis because of their soft and viscous mechanical behavior and because they are well shielded by bones. The viscoelastic properties of liver and brain have been the topic of several recent investigations related to diseases such as liver fibrosis (Asbach et al., 2008), hydrocephalus in brain (Taylor and Miller, 2004) and traumatic brain injury (Bayly et al., 2006). Further, MRE has been applied to liver and brain for the estimation and quantification of their viscoelastic properties (Klatt et al., 2007).

TABLE 1.2
Parameters of the Zener Model for Normal and Fibrotic Liver

Parameter		Unit	Liver Tissue Normal	Fibrotic
Viscosity	η	Pa s	7.3 ± 2.3	14.4 ± 6.6
Elastic Moduli	μ_1	kPa	1.16 ± 0.28	2.91 ± 0.84
	μ_2	kPa	1.97 ± 0.30	4.83 ± 1.77

MATLAB® PROGRAM FOR SIMULATING FREQUENCY-DEPENDENT COMPLEX MODULUS OF BRAIN TISSUE

```
close all
clear all
% enter frequency range %
fmin = 1;
fmax = 100;
% Enter model parameters %
% Voigit Model
uv = 1.81e3;etav = 2.9;
% Maxwell model
um = 2.28e3;etam = 15.1;
% Zener Model
uz1 = 0.84e3; uz2 = 2.03e3; etaz = 6.7;
% Jeffrey's model
etaj1 = 20.1; etaj2 = 0.9; uj = 2.23e3;
% Fractional Zener model
ufz1 = 0.58e3; ufz2 = 2.66e3; etafz = 7.8; alphafz =
0.85;
ep0 = 8.854e-12;
f = linspace(fmin, fmax, 100);
w = 2*pi*f;
j = sqrt(-1);
% Simulate Models
Gv = uv+(j*w*etav);
Gm = (j*w*etam*um)./(um+(j*w*etam));
Gz = ((uz1*uz2)+(j*w*etaz*(uz1+uz2)))./(uz2 + (j*w*etaz));
Gj = (-w*etaj1)*(((w*etaj2)-(j*uj))/
(uj+(j*w*(etaj1+etaj2))));
Gfz = ufz1+((ufz2*(((j*w*etafz)./(ufz2)).^alphafz)))./
(1+(((j*w*etafz)./(ufz2)).^alphafz));
% Plot
figure(1)
plot(f, real(Gv), f, real(Gm), f, real(Gz), f, real(Gj),
f, real(Gfz));
```

```
xlabel('frequency');ylabel('elastic modulus');
legend('voigt', 'maxwell', 'zener', 'jeffreys',
'fractional zener');
figure(2)
plot(f, imag(Gv), f, imag(Gm), f, imag(Gz), f, imag(Gj),
f, imag(Gfz));
xlabel('frequency');ylabel('viscous modulus');
legend('voigt', 'maxwell', 'zener', 'jeffreys',
'fractional zener');
```

The measures of soft tissue viscoelasticity are extremely useful in the diagnostics of soft tissue abnormalities as there is a significant variation in the mechanical properties of soft tissues in case of pathologies. For example, the mechanical modulus varies several times in the case of normal and cancerous soft tissues. Physicians normally use a technique called palpation to assess the conditions of soft tissues. In this technique, the physician feels the tissue using his fingers and assesses the stiffness or the elasticity of the tissue and makes diagnostic conclusions. This technique is entirely qualitative and relies upon the expertise of the medical doctor. However, a part of the tissue can be taken from the region under investigation and tested under standard conditions to assess or measure its elastic modulus and storage modulus using tensile or compression tests. In this case, only *ex-vivo* and *in-vitro* testing is possible and biopsy becomes a mandatory requirement, making the technique invasive and painful. If a method exists to measure the electrical properties of the soft tissue and derive or predict its mechanical properties, it would be non-invasive and *in-vivo* measurements could be made.

1.6 DIELECTRIC MODELS OF SOFT TISSUES

The electrical activity of the soft tissue can be modeled as a combination of resistors and capacitors, as shown in Figure 1.14.

Two major electrical properties of the soft tissue, namely, relative permittivity and conductivity, exhibit diagnostic information about tissue pathologies. Since the

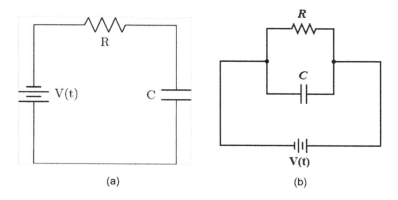

(a)   (b)

FIGURE 1.14   (a) RC series circuit, (b) RC parallel circuit.

beginning of the 20th century to the present day, the dielectric properties of biological soft tissues have been widely investigated to understand various phenomena involved with the interaction of electromagnetic fields with the soft tissues under analysis. By quantifying the dielectric properties of the tissues, it is possible to study the propagation of electromagnetic fields inside the tissues. The quantification of the dielectric properties of biological materials has a plethora of applications in biomedical and other industries (Lazebnik et al., 2006). Determination of the dielectric properties have applications in limiting human exposure to electromagnetic fields and in the design and development of technologies for the detection and treatment of cancer (Lazebnik et al., 2006).

The dielectric properties of tissues are determining factors of various types of soft tissue diseases such as cancers, tumors, fibrosis and cirrhosis. Knowledge of the complex permittivity of healthy and tumor tissues is crucial for diagnosing tissue pathologies during medical examinations and for bioelectromagnetic analysis of soft tissues (Zajíček et al., 2006). Similar to mechanical modulus, the dielectric permittivity is a complex variable and is a function of frequency. The dielectric permittivity of a tissue is expressed as:

$$\varepsilon(\omega) = \varepsilon'(\omega) + i\varepsilon''(\omega)$$

where the real part $\varepsilon'(\omega)$ is the relative permittivity and is expressed as a function of angular frequency ω, and the imaginary part $\varepsilon''(\omega)$ is the conductivity of the tissue, which is also a function of frequency ω. The variation of the relative permittivity of a biological soft tissue is often expressed using the dielectric relaxation models with three regions of dispersion or decay, namely, α, β, and γ regions at low, medium and high frequencies, respectively (Bone & Pethig, 1985), as shown in Figure 1.15.

The Debye model is a unique parameterized equation that models the variation of the relative permittivity and conductivity as a function of angular frequency. The Debye model is expressed as:

$$\hat{\varepsilon} = \varepsilon_\infty + \frac{\varepsilon_s - \varepsilon_\infty}{1 + j\omega\tau}$$

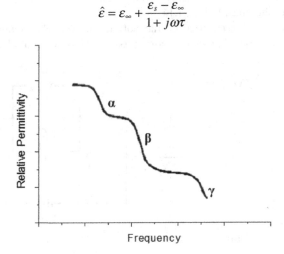

FIGURE 1.15   The α, β, and γ dispersions.

In this model, The electrical relaxation in tissues is expressed in terms of parameters, namely, the dielectric increment ε_s - ε_∞ and the relaxation time τ_n. In this expression, ε_∞ is the permittivity at field frequencies where, $\omega\tau \gg 1$, ε_s is the permittivity at $\omega\tau \ll 1$, and $j^2 = -1$. Broadening dispersion is possible by introducing a distribution parameter α_n, thus providing an alternative to the Debye model known as the Cole–Cole model (Gabriel et al., 1996).

The empirical equation of Cole and Cole (1941) was one of the pioneering models that provides a satisfactory depiction of the dielectric relaxation properties (Friedrich and Braun, 1992). Cole–Cole model is a parametric model that is commonly used as a compact mathematical representation of the wideband frequency-dependent dielectric properties. The variation of the dielectric properties of almost any soft tissue can be effectively predicted over a wide frequency range using the Cole–Cole model. The frequency-dependent dielectric property of the tissue is a complex quantity known as complex permittivity, which has a real part identified as the dielectric constant or relative permittivity and an imaginary part identified as the dielectric loss or conductivity. Analyzing these quantities over a wide range of frequencies is useful to better understand tissue behavior (Gabriel et al., 1996).

The Cole–Cole mathematical model provides an efficient and precise representation of dielectric property of various types of biological tissues such as liver, brain, prostate, fat, breast, skin and bone (Duck, 2013). The model can be used to predict the permittivity of soft tissues over a very wide frequency range.

$$\hat{\varepsilon}(\omega) = \varepsilon_\infty + \sum_n \frac{\Delta\varepsilon_n}{1 + (j\omega\tau_n)^{(1-\alpha_n)}} + \frac{\sigma_i}{j\omega\varepsilon_0} \tag{1.2}$$

where ω is the angular frequency, the real part is the frequency-dependent dielectric constant which is a measure of how much energy from an external electric field is stored in the material, the imaginary part is the frequency-dependent dielectric loss which is a measure of how lossy or dissipative a material is to an external electric field (Zajíček et al., 2006), n is the order of the Cole–Cole model, ε_∞ is the high frequency permittivity, τ_n is the relaxation time constant, $\Delta\varepsilon_n$ is the magnitude of the dispersion, σ_i is the static ionic conductivity and ε_0 is the permittivity of free space. Table 1.3 shows the parameters of the Cole–Cole model for soft tissues such as heart, liver, kidney and muscle, as presented by Gabriel et al. (1996).

Using the Cole–Cole model and the data available in the literature, the dielectric constant (relative permittivity) and dielectric loss (conductivity) of the human brain and liver tissue was obtained over a frequency range of 1–100 Hz, as presented in Figures 1.16a and b, respectively.

The measurement and mathematical modeling of the viscoelastic and dielectric properties are highly useful to understand the frequency-dependent behavior of the tissues in normal and abnormal conditions. Several pathological conditions such as fibrosis and tumors alter the material properties of the soft tissues. Hence, designing suitable measurement devices for localized measurement of such properties can offer new methods for the differential diagnosis and staging of various disorders related to tissues and organs of the human body.

TABLE 1.3
The Parameters of the Cole–Cole Model for Different Soft Tissues

Parameter	Heart	Liver	Kidney	Muscle
			Tissue	
ε_∞	4.0	4.0	4.0	4.0
$\Delta\varepsilon_1$	50.0	39.0	47.0	50.0
τ_1 (ps)	7.96	8.84	7.96	7.23
α_1	0.10	0.10	0.10	0.10
$\Delta\varepsilon_2$	12,000	6,000	3,500	7,000
τ_2 (ns)	159.15	530.52	198.94	353.68
α_2	0.05	0.20	0.22	0.10
$\Delta\varepsilon_3$	4.5×10^5	5.0×10^4	2.5×10^5	1.2×10^6
τ_3 (μs)	72.34	22.74	79.58	318.31
α_3	0.22	0.20	0.22	0.10
$\Delta\varepsilon_4$	2.5×10^7	3.0×10^7	3.0×10^7	2.5×10^7
τ_4 (ms)	4.547	15.915	4.547	2.274
α_4	0.00	0.05	0.00	0.00
σ_i	0.05	0.02	0.05	0.20

MATLAB® PROGRAM FOR SIMULATING FREQUENCY-DEPENDENT COMPLEX DIELECTRIC PROPERTIES OF BRAIN TISSUE

```
close all
clear all
fmin = 1;
fmax = 100;
ep0 = 8.854e-12;
f = linspace(fmin, fmax, 100);
w = 2*pi*f;
j = sqrt(-1);
einf = 4;
N = 4;
deleps = [45 400 2e5 4e7];
taun = [7.96e-12 15.92e-9 106.1e-6 5.305e-3];
alphan = [0.1 0.15 0.22 0.00];
sigma = 0.02;
epr = einf;
for n = 1:N,
epr = epr + deleps(n)./(1+(w*j*taun(n)).^(1-alphan(n)));
end;
epr = epr + sigma./(j*w*ep0);
```

```
k=-imag(epr).*w*ep0;
% Plot
figure(1)
loglog(f, real(epr), f, -imag(epr).*w*ep0);
xlabel('frequency');legend('relative permittivity',
'conductivity');
```

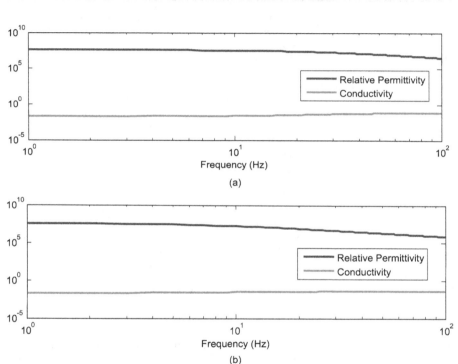

FIGURE 1.16   Variation of relative permittivity and conductivity of human (a) brain and (b) iver tissue.

REFERENCES

Ambikapathy, B. & Krishnamurthy, K. (2020). Mathematical modelling to assess the impact of lockdown on COVID-19 transmission in India: Model development and validation. *JMIR Public Health and Surveillance, 6*(2), e19368.

Arunachalam, K., Jacob, L. V., & Kamalanand, K. (2014). Design and analysis of finite element based sensors for diagnosis of liver disorders using biocompatible metals. *Technology and Health Care, 22*(6), 867–875.

Asbach, P., Klatt, D., Hamhaber, U., Braun, J., Somasundaram, R., Hamm, B., & Sack, I. (2008). Assessment of liver viscoelasticity using multifrequency MR elastography. *Magnetic Resonance in Medicine, 60*(2), 373–379.

Bailey, N. T. (1975). *The mathematical theory of infectious diseases and its applications.* Charles Griffin & Company Ltd, High Wycombe, Bucks.

Barnabas, R. V., Laukkanen, P., Koskela, P., Kontula, O., Lehtinen, M., & Garnett, G. P. (2006). Epidemiology of HPV 16 and cervical cancer in Finland and the potential impact of vaccination: Mathematical modelling analyses. *PLoS Medicine, 3*(5), e138.

Bayly, P. V., Black, E. E., Pedersen, R. C., Leister, E. P., & Genin, G. M. (2006). In vivo imaging of rapid deformation and strain in an animal model of traumatic brain injury. *Journal of Biomechanics, 39*(6), 1086–1095.

Bone, S. & Pethig, R. (1985). Dielectric studies of protein hydration and hydration-induced flexibility. *Journal of Molecular Biology, 181*(2), 323–326.

Caputo, M., Carcione, J. M., & Cavallini, F. (2011). Wave simulation in biologic media based on the Kelvin-Voigt fractional-derivative stress-strain relation. *Ultrasound in Medicine and Biology, 37*(6), 996–1004.

Chaplain, M. A. (2011). Multiscale mathematical modelling in biology and medicine. *IMA Journal of Applied Mathematics, 76*(3), 371–388.

Cole, K. S. & Cole, R. H. (1941). Dispersion and absorption in dielectrics I. Alternating current characteristics. *The Journal of Chemical Physics, 9*(4), 341–351.

De Pillis, L. G. & Radunskaya, A. (2001). A mathematical tumor model with immune resistance and drug therapy: An optimal control approach. *Computational and Mathematical Methods in Medicine, 3*(2), 79–100.

Dhar, P. R. & Zu, J. W. (2007). Design of a resonator device for in vivo measurement of regional tissue viscoelasticity. *Sensors and Actuators A: Physical, 133*(1), 45–54.

Duck, F. A. (2013). *Physical properties of tissues: A comprehensive reference book.* Cambridge, MA: Academic Press.

Edelstein-Keshet, L. (1988). *Mathematical models in biology* (Vol. 46). New Delhi: SIAM.

Erkamp, R. Q., Skovoroda, A. R., Emelianov, S. Y., & O'Donnell, M. (2004). Measuring the nonlinear elastic properties of tissue-like phantoms. *IEEE Transactions on Ultrasonics, Ferroelectrics, and Frequency Control, 51*(4), 410–419.

Findley, W. N. & Davis, F. A. (2013). *Creep and relaxation of nonlinear viscoelastic materials.* Chelmsford, MA: Courier Corporation.

Friedrich, C. & Braun, H. (1992). Generalized Cole–Cole behavior and its rheological relevance. *Rheologica Acta, 31*(4), 309–322.

Fung, Y. C. (2013). *Biomechanics: Mechanical properties of living tissues.* Berlin: Springer Science & Business Media.

Gabriel, S., Lau, R. W., & Gabriel, C. (1996). The dielectric properties of biological tissues: II. Measurements in the frequency range 10 Hz to 20 GHz. *Physics in Medicine and Biology, 41*(11), 2251.

Granich, R. M., Gilks, C. F., Dye, C., De Cock, K. M., & Williams, B. G. (2009). Universal voluntary HIV testing with immediate antiretroviral therapy as a strategy for elimination of HIV transmission: A mathematical model. *The Lancet, 373*(9657), 48–57.

Johnston, R. C., Brand, R. A., & Crowninshield, R. D. (1979). Reconstruction of the hip. A mathematical approach to determine optimum geometric relationships. *The Journal of Bone and Joint Surgery,* American volume, *61*(5), 639–652.

Kamalanand, K. & Jawahar, P. M. (2015a). Prediction of human immunodeficiency virus-1 viral load from CD4 cell count using artificial neural networks. *Journal of Medical Imaging and Health Informatics, 5*(3), 641–646.

Kamalanand, K. & Jawahar, P. M. (2015b). Comparison of swarm intelligence techniques for estimation of HIV-1 viral load. *IETE Technical Review, 32*(3), 188–195.

Kamalanand, K. & Jawahar, P. M. (2016). Comparison of particle swarm and bacterial foraging optimization algorithms for therapy planning in HIV/AIDS patients. *International Journal of Biomathematics, 9*(02), 1650024.

Kamalanand, K. & Jawahar, P. (2018). *Mathematical modelling of systems and analysis.* New Delhi: PHI Learning Pvt. Ltd.

Kamalanand, K. & Ramakrishnan, S. (2015). Effect of gadolinium concentration on segmentation of vasculature in cardiopulmonary magnetic resonance angiograms. *Journal of Medical Imaging and Health Informatics, 5*(1), 147–151.

Kamalanand, K. & Srinivasan, S. (2011a). Modelling and analysis of normal and atherosclerotic blood vessel mechanics using 3D finite element models. *ICTACT Journal on Soft Computing: Special Issue on Fuzzy in Industrial and Process Automation, 2*(1), 261.

Kamalanand, K. & Srinivasan, S. (2011b). Modeling of normal and atherosclerotic blood vessels using finite element methods and artificial neural networks. *World Academy of Science, Engineering and Technology, 60,* 1314.

Kamalanand, K., Sridhar, B. T. N., Rajeshwari, P. M., & Ramakrishnan, S. (2010). Correlation of dielectric permittivity with mechanical properties in soft tissue-mimicking polyacrylamide phantoms. *Journal of Mechanics in Medicine and Biology, 10*(02), 353–360.

Kamalanand, K., Srinivasan, S., & Ramakrishnan, S. (2011). Analysis of normal and atherosclerotic blood vessels using 2D Finite Element Models. *In 5th Kuala Lumpur International Conference on Biomedical Engineering* 2011 (pp. 411–414). Springer, Berlin, Heidelberg.

Kao, T. J., Saulnier, G. J., Isaacson, D., Szabo, T. L., & Newell, J. C. (2008). A versatile high-permittivity phantom for EIT. *IEEE transactions on Biomedical Engineering, 55*(11), 2601–2607.

Khoo, M. C. (2018). *Physiological control systems: Analysis, simulation, and estimation.* Hoboken, NJ: John Wiley & Sons.

Kiss, M. Z., Varghese, T., & Hall, T. J. (2004). Viscoelastic characterization of in vitro canine tissue. *Physics in Medicine and Biology, 49*(18), 4207.

Klatt, D., Hamhaber, U., Asbach, P., Braun, J., & Sack, I. (2007). Noninvasive assessment of the rheological behavior of human organs using multifrequency MR elastography: A study of brain and liver viscoelasticity. *Physics in Medicine and Biology, 52*(24), 7281.

Krishnamurthy, K. (2016). Parameter estimation of nonlinear biomedical systems using extended Kalman Filter algorithm: Development of patient specific models. *Computational Tools and Techniques for Biomedical Signal Processing, 76,* 24.

Krishnamurthy, K., Ambikapathy, B., Kumar, A., & De Britto, L. (2020). Prediction of the transition from subexponential to the exponential transmission of SARS-CoV-2 in Chennai, India: Epidemic nowcasting. *JMIR Public Health and Surveillance, 6*(3), e21152.

Krishnamurthy, K. & Jacob, L. (2014). Finite element based design of a capacitive sensor for diagnosis of cirrhotic and malignant liver. *Journal of Clinical and Experimental Hepatology, 4,* S65.

Krishnamurthy, K. & Ponnuswamy, M. J. (2018). Coupling of optimization algorithms based on swarm intelligence: An application for control of heroin addiction epidemic. *In Nature-Inspired Intelligent Techniques for Solving Biomedical Engineering Problems* (pp. 27–50). IGI Global, Pennsylvania.

Krishnamurthy, K., Sridhar, B. T. N., Rajeshwari, P. M., & Swaminathan, R. (2009). Correlation of electrical impedance with mechanical properties in models of tissue mimicking phantoms. *In 13th International Conference on Biomedical Engineering* (pp. 1708–1711). Springer, Berlin, Heidelberg.

Kumar, V. S., Anantharaj, U. J., Sakthioli, M., Vinnakota, R., & Krishnamurthy, K. (2016). Mathematical modelling of the effects of prebiotic concentration on lactobacillus casei growth. *International Journal of Infectious Diseases, 45,* 205.

Lazebnik, M., Converse, M. C., Booske, J. H., & Hagness, S. C. (2006). Ultrawideband temperature-dependent dielectric properties of animal liver tissue in the microwave frequency range. *Physics in Medicine and Biology, 51*(7), 1941.

Madsen, E. L., Zagzebski, J. A., & Frank, G. R. (1982). Oil-in-gelatin dispersions for use as ultrasonically tissue-mimicking materials. *Ultrasound in Medicine and Biology, 8*(3), 277–287.

Manickavasagam, K., Sutha, S., & Kamalanand, K. (2014). Development of systems for classification of different plasmodium species in thin blood smear microscopic images. *Journal of Advanced Microscopy Research, 9*(2), 86–92.

Pitkin, M., Raykhtsaum, G., Pilling, J., Shukeylo, Y., Moxson, V., Duz, V., ... & Prilutsky, B. (2009). Mathematical modeling and mechanical and histopathological testing of porous prosthetic pylon for direct skeletal attachment. *Journal of Rehabilitation Research and Development, 46*(3), 315.

Raphael, C. (1992). Mathematical modelling of objectives in radiation therapy treatment planning. *Physics in Medicine and Biology, 37*(6), 1293.

Rideout, V. C. (1991). *Mathematical and computer modeling of physiological systems* (p. 71). Englewood Cliffs, NJ: Prentice Hall.

Rowan, M. & Pedersen, P. (2006, October). P2C-3 an injury mimicking ultrasound phantom as a training tool for diagnosis of internal trauma. *In IEEE Ultrasonics Symposium*, 2006 (pp. 1612–1617). IEEE, Vancouver.

Schwarzer, G., Vach, W., & Schumacher, M. (2000). On the misuses of artificial neural networks for prognostic and diagnostic classification in oncology. *Statistics in Medicine, 19*(4), 541–561.

Stansfield, B. W., Nicol, A. C., Paul, J. P., Kelly, I. G., Graichen, F., & Bergmann, G. (2003). Direct comparison of calculated hip joint contact forces with those measured using instrumented implants. An evaluation of a three-dimensional mathematical model of the lower limb. *Journal of Biomechanics, 36*(7), 929–936.

Surry, K. J. M., Austin, H. J. B., Fenster, A., & Peters, T. M. (2004). Poly (vinyl alcohol) cryogel phantoms for use in ultrasound and MR imaging. *Physics in Medicine and Biology, 49*(24), 5529.

Takegami, K., Kaneko, Y., Watanabe, T., Maruyama, T., Matsumoto, Y., & Nagawa, H. (2004). Polyacrylamide gel containing egg white as new model for irradiation experiments using focused ultrasound. *Ultrasound in Medicine and Biology, 30*(10), 1419–1422.

Tanaka, G., Hirata, Y., Goldenberg, S. L., Bruchovsky, N., & Aihara, K. (2010). Mathematical modelling of prostate cancer growth and its application to hormone therapy. *Philosophical Transactions of the Royal Society of London A: Mathematical, Physical and Engineering Sciences, 368*(1930), 5029–5044.

Taylor, Z. & Miller, K. (2004). Reassessment of brain elasticity for analysis of biomechanisms of hydrocephalus. *Journal of Biomechanics, 37*(8), 1263–1269.

Watton, P. N., Luo, X. Y., Wang, X., Bernacca, G. M., Molloy, P., & Wheatley, D. J. (2007). Dynamic modelling of prosthetic chorded mitral valves using the immersed boundary method. *Journal of Biomechanics, 40*(3), 613–626.

Zajíček, R., Vrba, J., & Novotný, K. (2006). Evaluation of a reflection method on an open-ended coaxial line and its use in dielectric measurements. *Acta Polytechnica, 46*(5), 50–54

Zell, K., Sperl, J. I., Vogel, M. W., Niessner, R., & Haisch, C. (2007). Acoustical properties of selected tissue phantom materials for ultrasound imaging. *Physics in Medicine and Biology, 52*(20), N475.

Zhang, M., Castaneda, B., Wu, Z., Nigwekar, P., Joseph, J. V., Rubens, D. J., & Parker, K. J. (2007). Congruence of imaging estimators and mechanical measurements of viscoelastic properties of soft tissues. *Ultrasound in Medicine and Biology, 33*(10), 1617–1631.

2 Relationship between Viscoelastic and Dielectric Properties of Biological Soft Tissues

2.1 ELECTROMECHANICAL CORRELATIONS IN BIOLOGICAL SOFT TISSUES

The measurement of electrical and mechanical properties of soft tissues and organs are highly useful to analyze the physiological system and to classify normal and diseased tissues (Woo et al., 1981; Fatemi et al., 2003; Howe et al., 1995). Estimation of these properties of soft tissues is important as they are closely related to the tissue structure, biological conditions and pathology (Konofagou et al., 2004). Moreover, the properties of soft tissues are often investigated for the staging of several disease states (Duck, 2013; O'Rourke et al., 2007), as well as to understand and diagnose tissue pathologies (Fatemi et al., 2003; Asbach et al., 2008). Although it is known that the elastic modulus of soft tissues can vary as much as four orders of magnitude in healthy and diseased tissues (Duck, 2013; Sarvazyan, 1993), it is a very complicated biomechanical problem to measure the mechanical properties of living tissues (Novacek et al., 2002). A medical system that can predict or assess the mechanical properties of tissues could provide an important lead for the diagnosis of several soft tissue pathologies (Greenleaf et al., 2003).

The viscoelastic nature of soft tissues is described by the complex shear modulus, which has a real part known as the elastic or storage modulus and an imaginary part known as the viscous or loss modulus (Devi et al., 2007; Dasgupta & Weitz, 2005). The storage modulus and loss modulus are frequency dependent. The viscoelastic properties of biological tissue can be measured experimentally using various methods such as magnetic resonance elastography, ultrasound imaging methods, indentation devices and tactile sensors (Dhar & Zu, 2007). However, the cost of these measurement systems is high and *in-vivo* measurements are difficult to obtain. Moreover, some of the methods to measure the viscoelastic properties are highly invasive.

The dielectric properties of biological tissues have been a subject of active research as these properties fundamentally determine the propagation and interaction of electromagnetic fields within the tissue (Lazebnik et al., 2006). There is always significant variation in the dielectric properties of tissues in normal and abnormal cases (Lazebnik et al., 2007). For instance, the dielectric properties of malignant liver tissue is 19%–30% higher than normal liver tissue (O'Rourke et al., 2007).

The frequency-dependent dielectric property of the tissue is a complex quantity known as complex permittivity, which has a real part known as relative permittivity (dielectric constant) and an imaginary part known as conductivity (dielectric loss). Analyzing these quantities over a wide range of frequencies is useful to better understand tissue behavior (Gabriel et al., 1996; Foster & Schwan, 1989). The dielectric properties of biological tissues can be easily and efficiently estimated using electrical impedance spectrometers over a wide range of frequencies (Kun et al., 1999).

The interrelation and correlation between mechanical and electrical properties are intricate in case of soft tissues, and hence, the variations in mechanical properties may be estimated from electrical measurements (Kamalanand et al., 2010). Few evidence establish the interrelation between the electrical and mechanical properties of soft tissues from a biological viewpoint. The electrical and mechanical parameters are found to be interdependent in the case of cardiac tissues, cartilage, muscles and tendons (Whiteley et al., 2007; Youn et al., 2003). Keshtkar et al. (2008) assessed the urinary bladder volume changes on the bladder tissue impedance. Nash and Panfilov (2004) reported that a coupled electromechanical approach must be used in future modeling studies for the proper analysis of the physiological system. The dielectric permeability and conductivity of soft tissues were reported as a function of compressive strain by Chammas et al. (1994). Sierpowska et al. (2003) suggested that measurement of electrical properties of bone has the potential to provide the means for quantitative analysis of bone changes. The electrical and mechanical properties of tissue mimicking polyacrylamide phantoms were found to be closely correlated (Krishnamurthy et al., 2009; Kamalanand et al., 2010).

This chapter focuses on the correlations between the electrical and mechanical properties of soft tissues and intelligent models for the prediction of mechanical properties from electrical measurements. Furthermore, this chapter discusses two optimization algorithms for the estimation of the Cole–Cole model parameters from measurements of relative permittivity and conductivity as a function of frequency.

2.2 CORRELATION BETWEEN MECHANICAL AND ELECTRICAL PROPERTIES IN THE LIVER AND BRAIN TISSUE

The viscoelastic and dielectric properties of soft biological tissues need to be assessed for the accurate diagnosis of the physiological system. In the liver, these properties can be measured for the identification and staging of diseases such as hepatic malignancies and hepatic fibrosis (Asbach et al., 2008). Estimation of these properties over a wide range of frequencies enhances diagnostic performance by differentiating normal from abnormal tissues and classification of normal and abnormal tissues.

The objective here is to provide an extensive analysis of the correlations between the viscoelastic and dielectric properties of healthy liver and brain tissue using suitable dielectric and viscoelastic relaxation models. Figure 2.1 shows the variation of elastic or storage modulus as a function of relative permittivity in brain tissue. Non-linear relationships are observed between the frequency-dependent storage modulus estimated using spring-damper models such as the Maxwell model, Zener model, Jeffrey's model and the fractional Zener model, as well as the frequency-dependent relative permittivity estimated using the Cole–Cole model.

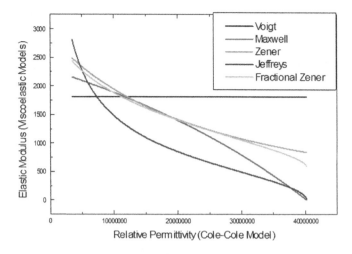

FIGURE 2.1 Co-variation of elastic modulus and relative permittivity in brain tissue.

Further, Figure 2.2 shows the non-linear relationships between the viscous modulus of brain tissue estimated using various spring-damper models and the conductivity of brain tissue estimated using the Cole–Cole model. There are strong positive correlations between the storage modulus and relative permittivity of brain tissue. Furthermore, there are strong negative correlations between the viscous modulus and the conductivity of brain tissue. Similar variations are observed in the case of liver tissue, as shown in Figures 2.3 and 2.4. Finally, the correlation between the frequency-dependent mechanical modulus (storage modulus and the viscous or loss modulus) and the frequency-dependent dielectric properties (relative permittivity and conductivity) are presented in Table 2.1.

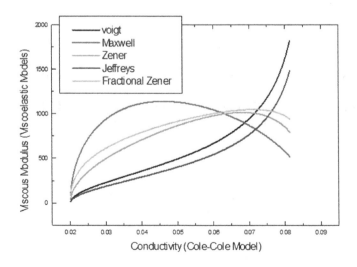

FIGURE 2.2 Co-variation of viscous modulus and conductivity in brain tissue.

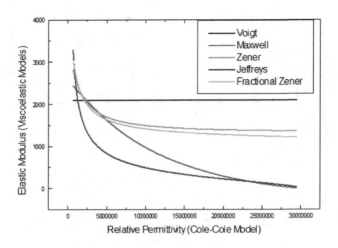

FIGURE 2.3   Co-variation of elastic modulus and relative permittivity in liver tissue.

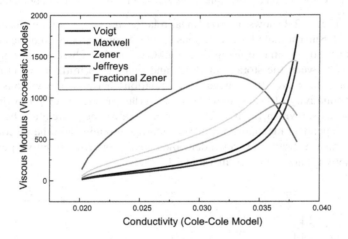

FIGURE 2.4   Co-variation of viscous modulus and conductivity in liver tissue.

2.3   ARTIFICIAL NEURAL NETWORKS IN THE PREDICTION OF SOFT TISSUE VISCOELASTIC PROPERTIES

A neural network is a model that can be constructed and trained in such a manner that the set of input variables generate the required set of output variables. Referring to this condition, the neural network is expressed as a set of inputs, $x_1, x_2, \ldots, x_n$, and a target dataset which is the required output of the network. The neuron output signal (out) is denoted by the following expression:

$$\text{out} = f(\text{net}) = f\left(\sum_{j=1}^{n}(x_j w_j)\right)$$

TABLE 2.1

Correlation Values between the Mechanical Modulus Estimated Using Different Models and the Complex Dielectric Properties Estimated Using the Cole–Cole Model of Brain and Liver Tissue

		Brain		Liver	
		Cole–Cole Model		Cole–Cole Model	
		Relative Permittivity	Conductivity	Relative Permittivity	Conductivity
Voigt Model	Storage Modulus	–	−0.9403	–	−0.7535
	Loss Modulus	–	0.9406	–	0.8144
Maxwell Model	Storage Modulus	−0.9946	0.1965	−0.9623	0.0567
	Loss Modulus	0.9946	−0.1972	0.9841	−0.1615
Zener Model	Storage Modulus	−0.9788	−0.8059	−0.7790	−0.9619
	Loss Modulus	0.9790	0.8054	0.8394	0.9593
Jeffereys Model	Storage Modulus	−0.9403	−0.9403	−0.7535	−0.7535
	Loss Modulus	0.9406	0.9406	−0.8144	0.8144
Fractional Zener Model	Storage Modulus	−0.9875	−0.8653	−0.7583	−0.9421
	Loss Modulus	0.9876	0.8650	0.8201	0.9707

where w_j is the weight matrix, and the function (f(net)) is known as the activation or transfer function and is expressed as:

$$f(\text{net}) = \frac{1}{1 + e^{\text{net}}}$$

During training, the output values generated by the network are compared with the target values, which are the required values of the outputs. Based on the error between the network outputs and the supplied target values, the weight matrix is updated using a suitable learning algorithm such as the back-propagation algorithm. Repeated weight updation is performed until the error is within required limits (Kamalanand and Jawahar, 2018).

In this chapter, different feed-forward back-propagation neural networks with a three-layered architecture is considered. The network was developed such that it could accept electrical measurements as inputs and could produce viscoelastic properties as outputs, as a function of frequency in the range of 1–100 Hz. The parameters of the neural network were estimated using a set of sample soft tissue properties consisting of the viscoelastic and electrical properties in the frequency range of 1–100 Hz, as the training set, which was adopted from the literature (Gabriel et al., 1996; Klatt et al., 2007). The results of three different networks with different input–output combinations are described.

The actual values of elastic and viscous modulus of the human brain tissue and the values predicted by the neural network using permittivity and conductivity as inputs are shown in Figure 2.5a and b, respectively. It is seen that the network can predict the values of elastic and viscous modulus in the adopted frequency range to a certain degree of accuracy. The Sum-Squared Error (SSE) in prediction of elastic and viscous modulus was found to be 0.284 and 0.546, respectively. Further, it appears that the developed model is more efficient in predicting the elastic modulus compared to the prediction of viscous modulus.

Similarly, Figure 2.6 shows the actual values of conductivity, elastic modulus and viscous modulus of sample brain tissue, as well as the values predicted by the Artificial Neural Network (ANN) model with only permittivity as the input variable in the frequency range of 1–100 Hz. It appears that the prediction error is less in the case of elastic modulus compared to conductivity and viscous modulus. The sum-squared error in prediction of conductivity, elastic modulus and viscous modulus was found to be 2.333, 0.402 and 3.657, respectively.

Finally, the actual and predicted values of permittivity, elastic modulus and viscous modulus of brain tissue, with conductivity as the input to the ANN model, are presented in Figure 2.7a–c respectively. It appears that the network can efficiently predict the values of elastic modulus when compared to the prediction of permittivity and viscous modulus. Further, the error in prediction of permittivity, elastic modulus and viscous modulus was found to be 2.441, 0.156 and 2.303, respectively. It appears that the prediction of elastic modulus of soft tissues from measurements of electrical properties is efficient in the frequency range of 1–100 Hz, using ANN.

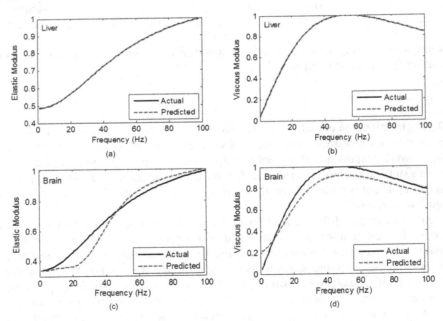

FIGURE 2.5 Actual and predicted values of elastic modulus and viscous modulus of liver (a and b) and brain tissue (c and d), with permittivity and conductivity as inputs to the ANN model.

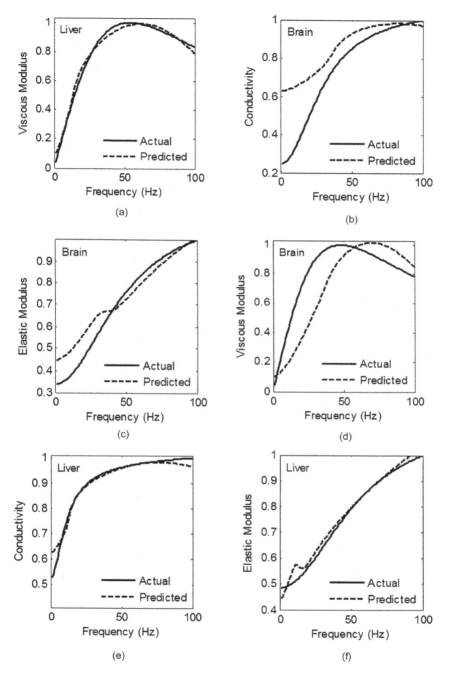

FIGURE 2.6 Actual and predicted values of conductivity, elastic modulus and viscous modulus of liver (a–c) and brain (d–f) tissue, with permittivity as inputs to the ANN model.

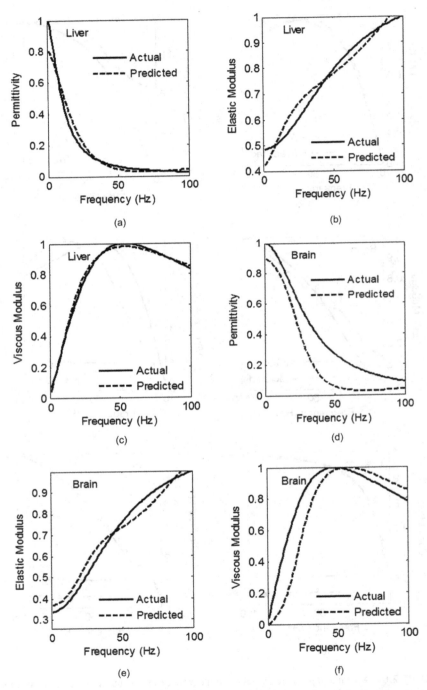

FIGURE 2.7 Actual and predicted values of permittivity, elastic modulus and viscous modulus of liver (a–c) and brain (d–f) tissue, with conductivity as an input to the ANN model.

Results demonstrate that the developed ANN models can predict the values of the elastic modulus to a high degree of accuracy compared to the other parameters. Further, it appears that the viscoelastic properties of biological soft tissues can be accurately estimated using electrical measurements. Such predictive models are useful as supportive noninvasive diagnostic tools for the identification and staging of soft tissue disorders.

2.4 PREDICTION OF SOFT TISSUE VISCOELASTIC PROPERTIES BY SUPPORT VECTOR REGRESSION

Consider a support vector regression (SVR) model in which the inputs are electrical properties such as the relative permittivity and conductivity. Using these inputs, the model predicts the values of the loss modulus and storage modulus and as a function of frequency.

SVR employs linear functions that are defined in a higher dimensional space for solving the regression problem. In SVR, the estimation is carried out by the minimization of risk using Vapnik's ε-insensitive loss function (Vapnik et al., 1996). SVM utilizes a risk function that consists of the empirical error and a regularization term which is obtained from the structural risk minimization principle. Given a set of inputs x_i, d_i as the required or the desired value and n as the total number of data patterns, SVMs approximate the function using the following equations (Tay & Cao, 2001):

$$y = f(x) = w\varphi(x) + b \tag{2.1}$$

where $\varphi(x)$ is the high-dimensional feature space which is nonlinearly mapped from the input space x. The coefficients w and b are estimated by minimizing the regularized risk function given by Equation (2.2).

$$R_{\text{SVMs}}(C) = C\frac{1}{n}\sum_{i=1}^{n} L_g\left(d_i, y_i\right) + \frac{1}{2}\parallel w \parallel^2 \tag{2.2}$$

In the work presented in this section, a radial basis kernel function (Gunn, 1998) is utilized to develop the SVR model.

$$\text{Radial Basis Kernel}: K(x, x') = e^{-\frac{\parallel x - x'\parallel}{2\sigma^2}} \tag{2.3}$$

where p_1 and p_2 are parameters specified by the user (Kamalanand & Jawahar, 2018).

This presents the results of two different SVR models with a radial basis function kernel of width = 3. In the first case, the SVR model was developed using the values of relative permittivity and conductivity as the inputs and the elastic modulus as the output of the model. Figure 2.8 shows the elastic modulus predicted by the SVR model and the actual values as a function of frequency. In the second case, the SVR model was developed using the frequency-dependent permittivity and conductivity

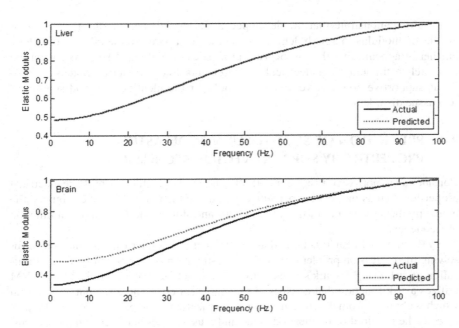

FIGURE 2.8 Variation of elastic modulus (actual and predicted values using the SVR model) shown as a function of frequency.

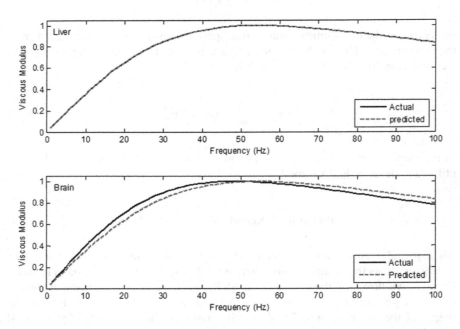

FIGURE 2.9 Variation of viscous modulus (actual and predicted values using the SVR model) shown as a function of frequency.

as inputs and the viscous modulus as the output of the model. The viscous modulus predicted by the SVR model and the actual values are shown as a function of frequency in Figure 2.9. It is seen that the SVR model is efficient in predicting the values of both elastic and viscous modulus with dielectric properties as the inputs.

In human physiological systems, the soft tissue electrical properties are well correlated with its mechanical properties. In several cases, due to the inaccessibility of tissue for measurement, it is not possible to measure the material properties effectively. The measurement of electrical properties of *in-vivo* soft tissues is easier compared to the mechanical measurements. Due to the existing correlations between the electrical and mechanical properties of the soft tissues, it is possible to estimate the unknown properties by measurement of few of the mechanical or the electrical properties using computational algorithms.

REFERENCES

Asbach, P., Klatt, D., Hamhaber, U., Braun, J., Somasundaram, R., Hamm, B., & Sack, I. (2008). Assessment of liver viscoelasticity using multifrequency MR elastography. *Magnetic Resonance in Medicine, 60*(2), 373–379.

Chammas, P., Federspiel, W. J., & Eisenberg, S. R. (1994). A microcontinuum model of electrokinetic coupling in the extracellular matrix: Perturbation formulation and solution. *Journal of Colloid and Interface Science, 168*(2), 526–538.

Dasgupta, B. R. & Weitz, D. A. (2005). Microrheology of cross-linked polyacrylamide networks. *Physical Review E, 71*(2), 021504.

Devi, C. U., Chandran, R. B., Vasu, R. M., & Sood, A. K. (2007). Measurement of visco-elastic properties of breast-tissue mimicking materials using diffusing wave spectroscopy. *Journal of Biomedical Optics, 12*(3), 034035.

Dhar, P. R. & Zu, J. W. (2007). Design of a resonator device for in vivo measurement of regional tissue viscoelasticity. *Sensors and Actuators A: Physical, 133*(1), 45–54.

Duck, F. A. (2013). *Physical properties of tissues: A comprehensive reference book.* Cambridge, MA: Academic Press.

Fatemi, M., Manduca, A., & Greenleaf, J. F. (2003). Imaging elastic properties of biological tissues by low-frequency harmonic vibration. *Proceedings of the IEEE, 91*(10), 1503–1519.

Foster, K. F. & Schwan, H. P. (1989). Electric properties of tissues and biological materials: A critical review CRC Crit. *Annual Review of Biomedical Engineering, 17*, 25–104.

Gabriel, S., Lau, R. W., & Gabriel, C. (1996). The dielectric properties of biological tissues: III. Parametric models for the dielectric spectrum of tissues. *Physics in Medicine and Biology, 41*(11), 2271.

Greenleaf, J. F., Fatemi, M., & Insana, M. (2003). Selected methods for imaging elastic properties of biological tissues. *Annual Review of Biomedical Engineering, 5*(1), 57–78.

Gunn, S. R. (1998). Support vector machines for classification and regression. *ISIS Technical Report, 14*(1), 5–16.

Howe, R. D., Peine, W. J., Kantarinis, D. A., & Son, J. S. (1995). Remote palpation technology. *IEEE Engineering in Medicine and Biology Magazine, 14*(3), 318–323.

Kamalanand, K. & Jawahar, P. (2018). *Mathematical modelling of systems and analysis.* New Delhi: PHI Learning Pvt. Ltd.

Kamalanand, K., Sridhar, B. T. N., Rajeshwari, P. M., & Ramakrishnan, S. (2010). Correlation of dielectric permittivity with mechanical properties in soft tissue-mimicking polyacrylamide phantoms. *Journal of Mechanics in Medicine and Biology, 10*(02), 353–360.

Keshtkar, A., Mesbahi, A., & Mehnati, P. (2008). The effect of bladder volume changes on the measured electrical impedance of the urothelium. *International Journal of Biomedical Engineering and Technology*, *1*(3), 287–292.

Klatt, D., Hamhaber, U., Asbach, P., Braun, J., & Sack, I. (2007). Noninvasive assessment of the rheological behavior of human organs using multifrequency MR elastography: A study of brain and liver viscoelasticity. *Physics in Medicine and Biology*, *52*(24), 7281.

Konofagou, E. E., Ottensmeyer, M., Agabian, S., Dawson, S. L., & Hynynen, K. (2004). Estimating localized oscillatory tissue motion for assessment of the underlying mechanical modulus. *Ultrasonics*, *42*(1–9), 951–956.

Krishnamurthy, K., Sridhar, B. T. N., Rajeshwari, P. M., & Swaminathan, R. (2009). Correlation of electrical impedance with mechanical properties in models of tissue mimicking phantoms. *In 13th International Conference on Biomedical Engineering* (pp. 1708–1711). Springer, Berlin, Heidelberg.

Kun, S., Ristic, B., Peura, R. A., & Dunn, R. M. (1999). Real-time extraction of tissue impedance model parameters for electrical impedance spectrometer. *Medical and Biological Engineering and Computing*, *37*(4), 428–432.

Lazebnik, M., Converse, M. C., Booske, J. H., & Hagness, S. C. (2006). Ultrawideband temperature-dependent dielectric properties of animal liver tissue in the microwave frequency range. *Physics in Medicine and Biology*, *51*(7), 1941.

Lazebnik, M., Popovic, D., McCartney, L., Watkins, C. B., Lindstrom, M. J., Harter, J., … & Temple, W. (2007). A large-scale study of the ultrawideband microwave dielectric properties of normal, benign and malignant breast tissues obtained from cancer surgeries. *Physics in Medicine and Biology*, *52*(20), 6093.

Nash, M. P. & Panfilov, A. V. (2004). Electromechanical model of excitable tissue to study reentrant cardiac arrhythmias. *Progress in Biophysics and Molecular Biology*, *85*(2–3), 501–522.

Novacek, V., Krakovsky, I., Muller, M., & Tonar, Z. (2002). Identification of mechanical parameters of biological tissues. *In Proceedings of the Conference Applied Mechanics* (pp. 267–272).

O'Rourke, A. P., Lazebnik, M., Bertram, J. M., Converse, M. C., Hagness, S. C., Webster, J. G., & Mahvi, D. M. (2007). Dielectric properties of human normal, malignant and cirrhotic liver tissue: In vivo and ex vivo measurements from 0.5 to 20 GHz using a precision open-ended coaxial probe. *Physics in Medicine and Biology*, *52*(15), 4707.

Sarvazyan, A. (1993). Shear acoustic properties of soft biological tissues in medical diagnostics. *The Journal of the Acoustical Society of America*, *93*(4), 2329–2330.

Sierpowska, J., Töyräs, J., Hakulinen, M. A., Saarakkala, S., Jurvelin, J. S., & Lappalainen, R. (2003). Electrical and dielectric properties of bovine trabecular bone: Relationships with mechanical properties and mineral density. *Physics in Medicine and Biology*, *48*(6), 775.

Tay, F. E. & Cao, L. (2001). Application of support vector machines in financial time series forecasting. *Omega*, *29*(4), 309–317.

Vapnik, V., Golowich, S. E., & Smola, A. J. (1996). Support vector method for function approximation, regression estimation and signal processing. *In* Advances in *Neural Information Processing Systems* (pp. 281–287). doi: 10.5555/2998981.2999021

Whiteley, J. P., Bishop, M. J., & Gavaghan, D. J. (2007). Soft tissue modelling of cardiac fibres for use in coupled mechano-electric simulations. *Bulletin of Mathematical Biology*, *69*(7), 2199–2225.

Woo, S. Y., Gomez, M. A., & Akeson, W. H. (1981). The time and history-dependent viscoelastic properties of the canine medial collateral ligament. *Journal of Biomechanical Engineering*, *103*(4), 293–298.

Youn, J. I., Akkin, T., & Milner, T. E. (2003). Electrokinetic measurement of cartilage using differential phase optical coherence tomography. *Physiological Measurement*, *25*(1), 85.

3 Electrophysiology

Physiology is a branch of biology that deals with the functions of human organs under normal conditions (Marieb & Hoehn, 2007; Tortora & Derrickson, 2018). Generally, the physiological systems are classified into ten different organ systems, as shown in Figure 3.1. All the physiological systems are interdependent and together perform the living process. From a systems perspective, the physiological systems are highly non-linear, extremely complex and multi-input and multi-output systems (Saladin, 2004; Martini, 2006). The measured physiological variables are not independent variables. There are several unmeasurable variables associated with the physiological system/process. The field of biomedical instrumentation is aimed at diagnosis and treatment of diseases and to support life or improve the quality of life (Khandpur, 1987). The objectives of biomedical instrumentation are (Singh, 2014):

- *Diagnosis*: To identify or determine the problem or the pathologies in the living system.
- *Monitoring*: To monitor or continuously assess the disease progression and the impact of the treatment.

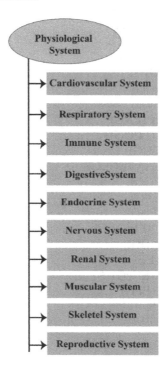

FIGURE 3.1 Physiological system.

- *Control*: Instruments aimed at treatment and control of pathologies, diseases or malfunctions.
- *Prosthesis*: Artificial body parts such as implants.

In recent years, biomedical instruments and artificial intelligence have been merged to improve the diagnostic accuracy and minimize the time and man hours, in particular, in high endemic situations. Several physiological measurable parameters such as both electrical and non-electrical quantities can be measured from physiological systems (Lessard, 2005). These measurements are well correlated with both the normal functioning as well as the pathologies associated with the physiological systems. The most common measurable non-electrical parameters associated with the physiological systems are acoustic, mechanical, chemical and optical signals. Electrophysiology is the field of study that explores the electrical activities in various cells, tissues and organs for the assessment of physiological and pathological states (Carter & Shieh, 2015). Measurable electrical parameters are action potential, resting potential and after potential from electrophysiological signals. Various electrophysiological signals originating from the human body are (Bachoud-Lévi, 2017):

- Electrocardiography (ECG/EKG)
- Electroatriography (EAG)
- Electroventriculography (EVG)
- Intracardiac Electrogram (EGM)
- Electroencephalography (EEG)
- Electrocorticography (ECOG)
- Electromyogram (EMG)
- Electrooculography (EOG)
- Electroretinography (ERG)
- Electronystagmography (ENG)
- Electroantennography (ECOG)
- Electrogastrography (EGG)
- Electrogastroenterography (EGEG)
- Electroglottography (EGG)
- Electropalatography (EPG)
- Electroarteriography (EAG)
- Electroblepharography (EBG)
- Electrodermography (EDG)
- Electrohysterography (EHG)
- Electroneuronography (ENeG)
- Electropneumography (EPG)
- Electrospinography (ESG)
- Electrovomerography (EVG)

Electrophysiology techniques support the clinicians and researchers to explore the intercellular and intracellular messages (Carter & Shieh, 2015).

3.1 ACTION POTENTIALS

3.1.1 BIOPOTENTIAL

The human physiological system comprises mechanical, electrical and chemical systems that facilitate our daily activities (Ha et al., 2014; Van Drongelen, 2018). For example, the mechanical system of the muscles contract using actin and myosin filaments present in the muscles. The chemical system in the body enables the neurotransmitter for inter-cell communication, and the electrical system allows the generation and propagation of the electrical potentials from nerve cells and muscle fibers. Such electrical potentials are responsible for the movement of muscle along with the various organs such as the ophthalmic, cardiac, sensory and brain function. The electrical potential differences are created by the flow of electrical ions in and out of cells. These electrical potential differences are referred as biopotentials.

Biopotentials are generated as a result of the electrochemical activity of the cells that are the components of the nervous, muscular and granular tissue. The electrical activity of the cell is generated via in and out ion movement (K^+, Na^+ and Cl^-) through the cell membrane (Van Drongelen, 2010). Generally, the potentials are in two states (active and resting potential). The active potential is generated when cells are stimulated. The membrane potential when the cell is inactive is the resting potential. At resting state, the potassium ion is more permeable in the cell membrane when compared to the sodium and potassium ion concentrations is higher in the interior of the cell when compared to the exterior of the cell. The diffusion gradient of potassium ion arises toward the exterior of the cell that creates more negative ions in the interior of the cell. At steady or depolarization state, the diffusion gradient of potassium ion is in equilibrium and balanced by the electric field with the polarization voltages of −70 mV (Thakor, 2015; Webster, 1984). If the cells are electrically stimulated, the diffusion gradient of potassium ion increases and diffuse toward the interior of the cells that creates more potential. If the active potentials reach +40 mV, the permeability of the potassium ion decreases and the sodium ion increases causing resting potential. This cycle produces the several cellular potentials called as action potentials (Yazıcıoğlu et al., 2009; Thakor, 2015).

3.1.2 PROPAGATION OF ACTION POTENTIALS

Action potentials are the electrical signals that accompany the mechanical contraction of a single cell when stimulated by an electric current (Hammond, 2014). Action potentials are caused by flow of ions across the cell membrane (White, 2002; Hammond, 2014) which provide information on the anatomical structure and physiological functions of the cell. Figure 3.2 shows the typical action potential along with its various parts.

At resting or polarized state, the negative potassium ions are more in the interior of the cell membrane when compared to the exterior of the cell membrane. The ion exchange does not occur in the interior as well as exterior of the cell membrane (McCormick, 2014; Hammond, 2014). In the depolarized state, due to the stimulation of the cell, the positive ions penetrate into the cell membrane, generating action

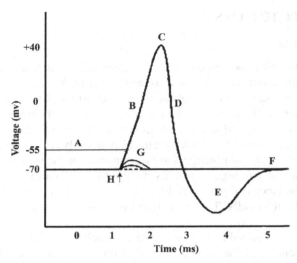

A - Threshold; B - Depolarization; C - Action potential; D - Repolarization;
E - Refractory period; F - Resting State; G - Failed Initiations; H - Stimulus

FIGURE 3.2   Action potential.

potentials. During the repolarization state, the potentials attains the equilibrium and it reaches to resting potentials (Hammond, 2014).

According to the all-or-nothing law, the magnitude of the action potential is the same, irrespective of the type and strength of stimulus.

3.2 HODGKIN–HUXLEY MODEL

Sir Alan Hodgkin and Sir Andrew Huxley developed the principal mathematical framework for modern biophysical neural circuit modeling. In 1940s and 1950s, Hodgkin and Huxley performed the consecutive electrophysiological experiments on the squid giant axon using *longfin inshore squid* (Koslow & Subramaniam, 2005; Maršálek, 2000). The squid giant axon is approximately 0.5 mm in diameter and provides the benefit of insertion of voltage-clamp electrodes inside the lumen of the axon. Further, Hodgkin and Huxley demonstrated the changes in the conductance of Na^+ and k^+ in the axon membrane through the ionic current movements in the squid giant axon (Horikawa, 1998; Koslow & Subramaniam, 2005). Hodgkin and Huxley developed a mathematical model of the electric potentials and the time-dependent properties of the conductance of Na^+ and k^+ ions based on the consecutive voltage-clamp experiments. The developed model describes the biophysical properties of the action potential, which is known as the Hodgkin–Huxley model (Koslow & Subramaniam, 2005).

In physiological system neural modeling, the electrical equivalent circuit of a neuron are developed using the electrical properties of the neuron. Figure 3.3 shows the equivalent electrical circuit to describe the electrical activity in a squid giant axon.

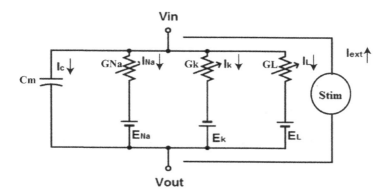

FIGURE 3.3   Electrical equivalent circuit for a squid giant axon.

The capacitor C_m represents the storage capacity of the cell membrane. G_{Na} and G_k are the two independent variable resistors which represents the voltage-dependent Na+ and K+ conductance. G_L represents the voltage-independent leakage conductance. *Stim* represents the externally stimulated current. Current across the cell membrane has two major components, namely, the membrane capacitance and flow of ions through the resistive membrane channel (Koslow & Subramaniam, 2005). I_c is the capacitive current and defines the rate of change of charge q at membrane channels (Hodgkin & Huxley, 1952).

$$I_c = \frac{dq}{dt} \tag{3.1}$$

where $q = C_m V_m$. C_m is the membrane capacitance, V_m is the membrane voltage and then the capacitive current is rewritten as:

$$I_c = C_m \frac{dV_m}{dt} \tag{3.2}$$

The current associated with flow of ions through resistive membrane channels is expressed as:

$$I_{ion} = I_{Na} + I_K + I_L \tag{3.3}$$

The three ionic currents are presented in this model, sodium current I_{Na}, potassium current I_K and small leakage current due to chloride ions I_L. The general differential equation of the electrical circuit is given by Hodgkin and Huxley (1952):

$$C_m \frac{dV_m}{dt} + I_{ion} = I_{ext} \tag{3.4}$$

where I_{ext} is the external stimulus current from outside to inside. I_c, I_{Na}, I_K and I_L are the ionic currents from inside to outside. According to Ohms law:

$$V = IR$$

$$I = \frac{V}{R}, \frac{1}{R} = G \tag{3.5}$$

$$I = GV$$

where G is the conductance. In this model, E_{Na}, E_K and E_L are equilibrium potentials at which the total ionic current passing across the membrane would be zero. Hence, the total ionic current is given by Koslow & Subramaniam (2005) and Hodgkin & Huxley (1952),

$$I_{ion} = \sum_j G_j \left(V_m - E_j \right) \tag{3.6}$$

Or

$$I_{ion} = G_{Na} \left(V_m - E_{Na} \right) + G_K \left(V_m - E_K \right) + G_L \left(V_m - E_L \right) \tag{3.7}$$

3.3 BIOMEDICAL SIGNALS

Such health information is acquired by physical measurement equipment that measures temperature, oxygen concentration, heart rate, blood pressure, blood glucose level, electrophysiological signals, etc. The biomedical system generates signals that are used to describe the physiological phenomenon of the human body (Chua et al., 2010). The biomedical signal represents the useful biological information about the dynamic activity of the physiological system. The commonly measured biomedical signals are electroencephalograms, electrocardiograms, carotid pulse signal, speech signals and phonocardiogram signals (Mitov, 1998; James & Hesse, 2004). The five sources of noises that influence the acquisition of biosignals are thermal noise, aliasing, instrument noise, power line alternating current and interference (Bruce, 2001; Jung et al., 2000). There are five different categories of biosignals which are classified based on their origin, biosignal model, biosignal nature, dimensionality and number of channels (Bruce, 2001; Penzel et al., 2001; Majdalawieh et al., 2003).

3.3.1 CLASSIFICATION BASED ON THE BIOSIGNAL SOURCE

Biosignals that originate from different sources of physiological systems such as nervous system, endocrine system, auditory system, vision system, respiratory system, musculoskeletal system, gastrointestinal system, circulatory system and cardiovascular system (Akay, 1996; Cattani et al., 2012).

3.3.2 CLASSIFICATION BASED ON THE BIOSIGNAL MODEL

Biosignals can be either be stochastic or deterministic. Further, deterministic signals can be either periodic or non-periodic biosignals (Naït-Ali, 2009; Semmlow & Griffel, 2014).

3.3.3 CLASSIFICATION BASED ON THE DIMENSIONALITY

Biosignals can be one-dimensional (electrocardiogram signal, electromyogram signal, etc.) or two-dimensional (2D images such as X-ray images) or multidimensional (magnetic resonance images, functional magnetic resonance images, etc.) (Semmlow & Griffel, 2014; Akay, 1996).

3.3.4 CLASSIFICATION BASED ON THE NUMBER OF CHANNELS

Based on the number of channels for acquiring biosignals, they can be single-channel signals such as the pulse waves of multi-channel to measure signals such as the EEG signals (Semmlow & Griffel, 2014; Bhoi et al., 2015).

3.3.5 CLASSIFICATION BASED ON THE NATURE OF THE BIOSIGNALS

Based on the nature of biosignals, they are categorized into different types such as optical, electrical, mechanical, magnetic, chemical, mechanical and thermal types (Kaniusas, 2012; Semmlow & Griffel, 2014).

- Chemical biosignals are the signals that provide information about the concentration of different chemicals in the human body (pH, oxygen, level of glucose, etc.).
- Optical biosignals are produced either naturally or induced by the light or optical attributes of the physiological systems.
- Electrical signals are generated by the potential difference associated with the nerve and muscle tissue.
- Mechanical biosignals are produced by the mechanical activities of the biological process such as motion, displacement and pressure (blood pressure measurement).
- Acoustic biosignals are produced from the respiratory sounds, cough, wheeze and snoring sounds.

3.4 PROBLEMS ASSOCIATED WITH RECORDING BIOMEDICAL SIGNALS

Although the measurement of biosignals provide highly useful information about the physiological system, there are several problems associated with the measurement. These problems need to be addressed in each stage so that the diagnostic information acquired using the measurement is correct and valid. The common problems associated with the measurement include:

- The effect of instrumentation or procedure on the system
- Access of the variable to the measurement
- Variability of signal source (dynamic, random, periodic signals)
- Artifacts such as breathing, motion, cross-talk, ambiants, etc.
- Energy limitations
- Patient safety (shock and radiation hazards)

3.5 PHYSIOLOGICAL TRANSDUCERS

A physiological transducer converts one form of energy associated with the physiological system into an electrical signal. The quantities such as blood pressure, blood flow in the vessel, respiratory airflow, body movements, temperature, electric potential, biomagnetism and body fluid control is measured using the measurement systems based on physiological transducers (Brown & Gupta, 2008; Macy, 2015). Generally, there are two types of transducers, namely, input and output transducers. The input transducer is also called a sensor and is used to convert different kinds of physical energies (temperature, pressure, sound, acceleration) into electrical energy. The output transducer is used to convert electrical energy into other physical quantities (Togawa et al., 1997; Macy, 2015).

The input transducer is of two different types, namely, active and passive transducers. Active transducers convert physical quantities into electrical signals. The examples of active sensors are piezo-electric devices, solar cells and thermocouples. Passive transducers are used to convert the changes in the physical variables into variations in capacitance, inductance and resistance.

3.5.1 EXAMPLE OF BIOMEDICAL TRANSDUCERS

The mechanoelectrical, magnetic induction and Doppler ultrasound are the mechano-electric sensors/transducers that are used to detect the mechanical forms of physical quantities, such as motion, pressure, velocity, mechanical power, volume and flow of fluids (Kozlíková, 2010). Thermometers, liquids crystals and calorimeter are thermal sensors that are used to detect the body core temperature, surface temperature distribution and emission of infrared radiation. The oximeter and glucometer are biochemical sensors that are used to detect the pH, glucose concentration and partial pressure of gasses. The electrodes act as a sensor to acquire the action potentials and display the electrophysiological signals such as ECG, EMG and EEG (Togawa et al., 1997).

3.6 ELECTRODE TYPES

In the physiological system, the ions act as charge carriers and whose movement provide the key mechanism for electric conductivity in the human body (Albulbul, 2016; Enderle & Bronzino, 2012). Consequently, the electrodes and the measurement system are required to collect the electrical signals from the localized parts of the human body. The electrical signals from the body are measured using electrodes which are made up of electrical conductors and conducting fluids. The interaction between the ions in the body and electrons in electrodes can significantly affect sensor performance. These reactions can be described for movement of charge in between the electrons and ions using the Equations 3.8 and 3.9 (Bronzino, 2000; Enderle & Bronzino, 2012):

$$C \rightleftharpoons C^{n+} + ne^- \tag{3.8}$$

$$A^{m-} \rightleftharpoons A + me^- \tag{3.9}$$

where C *and* A is the cation and anion material; n and m is the valency of the cation and anion material. In these reactions, the oxidation and reduction process are carried out continuously for charge movement of electrons and ions.

The concentration of the ions in the solution changes during the electron and ion interaction. Hence, the charge neutrality is inconsistent in this region, and hence, the electrolyte surrounding the conducting metal is at a different electrical potential from the rest of the solution. This potential difference is called a half-cell potential (Bronzino, 2000). The half-cell potential is most important when using the electrodes for DC measurement (Albulbul, 2016; Bronzino, 2000).

In electrochemistry, the interrelationship between the electric and ionic activities are commonly considered for electrode design. Using the Nernst Equation (3.10), the electric potential (E) will be present between the ions on either side of the cell membrane (Bronzino, 2000):

$$E = -\frac{RT}{nF}\ln\left(\frac{a_1}{a_2}\right) \tag{3.10}$$

where T is the absolute temperature, R is the universal gas constant, n is the valency of ions, F is the Faraday constant and a_1 and a_2 are the activities of the ions.

The idealized electrodes are categorized into two types based on the flow of current through the electrode, namely, polarizable and non-polarizable electrodes (Albulbul, 2016). The polarizable electrodes perform like capacitors because only transient current flows through the electrode or electrode junction. The non-polarizable electrodes perform like resistors because the non-transient current flows through the electrode junction. At equilibrium potential, the potential of the electrode does not change in the case of non-polarized electrodes when the large current passes through the electrode junction (Macy, 2015). Electrode potentials are changed in the case of polarized electrodes when a small current passes through the electrode junction. Hence, the electrode reaction occurs quickly in the case of non-polarized electrodes compared to the polarized electrodes. The total potential (V_t) comprises equilibrium potential (V_e) and overpotential (V_p) (Macy, 2015).

$$V_t = V_e + V_p \tag{3.11}$$

where $V_p = V_r + V_a + V_c$. The overpotential (V_p) is the difference between the potential at equilibrium condition and the operating states when the current is flowing in an electrode. There are three elements in the overpotentials, namely, resistive or Ohmic overpotential (V_r), activation overpotential (V_a) and concentration overpotential (V_c).

In the biological signal acquisition, non-polarized electrodes are preferred for low-frequency measurements, whereas polarized electrodes are used for transient electrical stimulation at higher frequency range. Depending on the size and class of the electrodes, they are broadly classified into five types, namely, microelectrodes, needle electrodes, surface electrodes, fine wire electrodes and electrode arrays (Macy, 2015).

Microelectrodes consist of an ultrafine-tapered tip and are used for either signal acquisition in single cells or electrical stimulations of nerve tissues (Plieth, 2008). The electrode tip must be small when compared to the cell dimension to avoid cell damage and to enable easy penetration into the cell wall. There are three categories of microelectrodes namely, glass micropipettes, metal microelectrodes and solid-state microprobes (Enderle & Bronzino, 2012), as shown in Figure 3.4. Mostly, solid-state microelectrodes are used for multi-channel recordings of biopotential or electrical stimulation of neuron cells in brain or spinal cord (Enderle & Bronzino, 2012; Manahan-Vaughan, 2018). The major advantage of using solid-state microelectrodes instead of other two microelectrode types is the capability to manufacture small size electrodes in a mass quantity.

Fine wire electrodes are referred as intra-muscular electrodes which used to record and extract the information about the activity of the particular muscle region (Rudroff, 2008). Fine wire electrodes consist of a pair of fine nylon-coated wire with diameter of around 50 μm or less with unisolated ending, which looks like a hypodermic needle, with a small barb at the end of the wires to keep them in the tissue, as shown in Figure 3.5. These electrodes have 25 or 27-gauge sharp-edge needles to reduce the pain and can easily be inserted into the muscle without the use of anesthetics.

Needle electrodes are also referred as intramuscular wire electrodes which can be directly inserted into the muscle tissue to record the action potential more accurately (Rubin, 2019). The needle electrodes are classified into four types, namely,

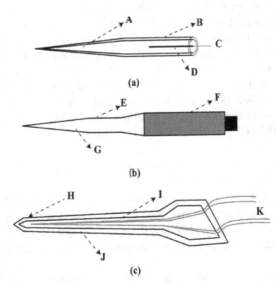

A - KCl electrode; B - Glass Capillary, C - Lead; D - Ag/AgCl wire; E - Metal; F - Insulation; G - Shank;
H - Exposed Recording Electrodes; I - Insulated Leads; J - Support Structure; K - Leads

FIGURE 3.4 Microelectrodes (a) glass micropipettes, (b) metal microelectrodes and (c) solid-state microelectrodes.

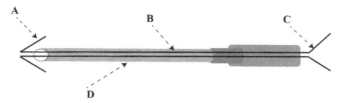

A - Unisolated Ending
B - Steel Cannula
C - Un-isolated Ending Electrode Site
D - Hooked Electrode Wire

FIGURE 3.5 Fine wire electrodes.

concentric needle electrodes, monopolar needle electrodes, single fiber-electrodes and macro-electrodes (Subasi, 2019), as shown in Figure 3.6. The monopolar needle consists of an insulated metal except on the needle tip and is made up of stainless steel with gauge ranging from 26 to 31. The tip is conical and the conductive area of the needle is around 0.10–0.50 mm². The concentric needle electrodes consist of a reference electrode (cannula) made up of insulated metal and the active electrode (core) made up of silver. The diameter of cannula is around 0.45–0.7 mm and the diameter of the core is 0.1 mm and tip of the needle has a flat elliptical shape around an angle of 15°. The exposure area of the concentric needle electrodes is around 0.02–0.1 mm² (Holstege et al., 2014). Single fiber needle electrodes consist of a small recording surface around 25 μm and exposure of the needle into the muscle is 3 mm. The single fiber needle electrodes are most sensitive in detecting neuromuscular disorders compared to the other electrodes (Selvan, 2011). The macro needle electrode consists of a 50-mm steel cannula and the tip of the electrodes is the platinum wire of 25 μm in diameter (Stålberg, 1980).

Surface electrodes are electrodes which can be positioned on the surface of the skin either using adhesive tape or adhesive electrode gel (Jamal, 2012; Zhang & Hoshino, 2018). It consists of conductive materials such as metal alloy, metals, metal compounds, and conductive rubber, as shown in Figure 3.7. The surface electrodes are widely used to acquire biosignals noninvasively. They are broadly classified into

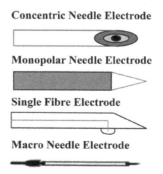

FIGURE 3.6 Needle electrodes.

FIGURE 3.7   Surface electrodes.

four types, namely, metal plate electrodes, floating electrodes, suction cup electrodes, adhesive-type electrodes and multi-point-type electrodes. The metal plate electrodes are a better sensitive-type electrodes which consist of either rectangular or circular shaped electrodes made of silver, nickel or German silver materials. The electrode has a small contact area and is placed on the surface of the skin using the electrolyte paste. The suction cup electrodes are made up of plastic syringe barrel, suction tube and cables. The suction cup electrodes are attached to the flat surface of the body and needs high contact impedance. The maintenance of these electrodes is difficult due to infections and cleaning procedures. Adhesive electrodes are light in weight and have a pad for applying the electrode paste. The multi-point-type electrodes consist of more than 1,000 active points. The floating electrodes are placed in contact with the surface of the skin using the electrolyte jelly or paste instead of direct contact on the body and electrodes are used to avoid movement errors and to maintain the mechanical stability. Electrode arrays are organized in the form of a grid with different shape, and the size of the array ranges from micrometer to centimeters in the surface area (Macy, 2015).

3.7 SKIN–ELECTRODE MODEL

In an electrophysiological signal acquisition, electrode performance is extremely dependent on the impedance of the electrode–skin interface. The impact of the high electrode–skin impedance in the measurement of biological signals leads to poor signal quality, low signal-to-noise ratio and low signal amplitude (Albulbul, 2016). The proper selection of electrodes that have low electrode–skin impedance is necessary for biological measurements. Figure 3.8 shows the equivalent circuit model for the electrode–skin interface.

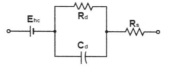

FIGURE 3.8 Equivalent circuit model for electrode–skin interface.

The impedance of the electrode–skin interface for a single electrode is expressed as (Albulbul, 2016):

$$Z_e = R_s + \frac{R_d}{1 + j2\pi f C_d R_d} \tag{3.12}$$

where R_d is the resistance that may occur due to the charge transfer between the skin and the electrode, R_s is the series resistance of electrolyte gel and sweat, E_{hc} is the potential difference between the skin and electrode, C_d is the capacitance between the electrode and skin layer and f is the frequency (Hz) (Albulbul, 2016).

3.8 TYPICAL SIGNALS

3.8.1 ELECTROCARDIOGRAM (ECG)

The cardiopulmonary system consists of blood vessels that carry nutrients and oxygen to the tissues and removes carbon dioxide from the tissues in the human body (Humphrey & McCulloch, 2003; Alberts et al., 1994). Blood is transported from the heart through the arteries and the veins transport blood back to the heart. The heart consists of two chambers on the top (right ventricle and left ventricle) and two chambers on the bottom (right atrium and left atrium). The atrioventricular valves separates the atria from the ventricles. Tricuspid valve separates the right atrium from the right ventricle, mitral valve separates the left atrium from the left ventricle, pulmonary valve situates between right ventricle and pulmonary artery, which carries blood to the lung and aortic valve situated between the left ventricle and the aorta which carries blood to the body (Bronzino, 2000). Figure 3.9 shows the schematic diagram of heart circulation and there are two components of blood circulation in the system, namely, pulmonary and systemic circulation (Humphrey, 2002; Opie, 1998; Milnor, 1990). In pulmonary circulation, pulmonary artery transports blood from heart to the lungs. The blood picks up oxygen and releases carbon dioxide at the lungs. The blood returns to the heart through the pulmonary vein. In the systemic circulation, aorta carries oxygenated blood from the heart to the other parts of the body through capillaries. The vena cava transports deoxygenated blood from other parts of the body to the heart.

Cardiovascular disorders affect the blood vessels and heart in the physiological system (Serhani et al., 2020). According to the World Health Organization (WHO), 17.9 million people (31%) die each year due to cardiovascular diseases (WHO report). According to the WHO cardiovascular disorders are a leading cause of death, in particular, heart attacks in middle-aged groups. The most cost-effective electrodiagnostic method for the detection of the cardiovascular disorders is the ECG (Hassan et al., 2019). ECG is the process of measuring the electrical activity of the heart

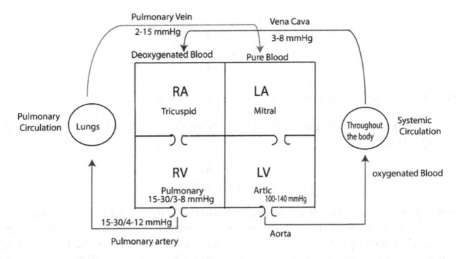

FIGURE 3.9   Schematic diagram of heart circulation.

muscles and these electrical signals are referred as electrocardiograms (Preejith et al., 2016). These electrical signals are plotted as a graph of voltage with respect to time, measured using the electrodes placed on the surface of the skin, at specific locations (Ghosh et al., 2014; De Cooman et al., 2014). The electrodes acquire electrical variations in the heart muscles, and such recorded ECG signals can be used to diagnose several cardiac abnormalities such as disturbances in cardiac rhythm, atrial fibrillation, ventricular tachycardia, myocardial ischemia, myocardial infarction, hypokalemia and hyperkalemia. Generally, 10 electrodes are placed on the surface of the chest area, the upper and lower limbs and the magnitudes of the electrical signals are recorded. In ECG signals, the P wave is correlated with the depolarization of the atria, QRS complex is correlated with the depolarization of the ventricles and T wave is correlated with the repolarization of the ventricles (Gierałtowski et al., 2014; Marston et al., 2019). The pattern of the ECG signals is unique and same for all normal individuals. Figure 3.10 shows a typical electrocardiogram signal for 10 s, available at the opensource database (https://physionet.org/) and a typical ECG wave.

3.8.2   ELECTROMYOGRAMS

In the physiological system, the neuromuscular system consists of the muscular system and the nervous system (Davis & Cladis, 2016; Reed et al., 2017; Begg et al., 2007). The neuromuscular system is a highly complex electro-mechanical system which includes various subsystems with interconnected variables (Smelser & Baltes, 2001; Roberg & Roberts, 1996). The neuromuscular disorders affects primarily the nervous system that has the direct impact on the muscles (Blottner & Salanova, 2015). Several disorders, such as myopathies, amyotrophic lateral sclerosis, multiple sclerosis, myasthenia gravis (Greenway et al., 2006; Rowland & Shneider, 2001; Deenen et al., 2015; Blottner & Salanova, 2015), etc., affect the muscular as well as the nervous system, which results in increased mortality rate and decreased quality of life.

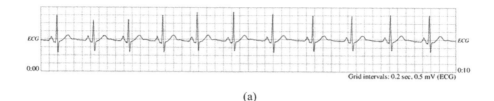

(a)

QRS
Complex

R

PR
Segment

ST
Segment

P

T

PR Interval

Q

S

QT Interval

(b)

FIGURE 3.10   (a) Typical electrocardiogram signal and (b) typical ECG wave.

EMG (Sadikoglu et al., 2017) is an electrodiagnostic technique and the process of recording the electrical activity of the muscle contraction. The recorded signals are used for the analysis and diagnosis of various neuromuscular disorders, such as myopathy, multiple sclerosis, peripheral neuropathy, amyotrophic lateral sclerosis (Blottner & Salanova, 2015; Duque et al., 2014; Greenway et al., 2006; Rowland & Shneider, 2001), etc. Generally, the nominal frequency range of the measured EMG signal is between 0 to 500 Hz and the frequency range for healthy muscles is 30–60 Hz (Acharya et al., 2011). Figure 3.11 shows the typical EMG signals for 10 s, available at the opensource database (https://physionet.org/).

FIGURE 3.11   Typical EMG signal.

3.9 BIOSIGNAL AMPLIFIERS

The amplitude of electrophysiological signals such as EMG and ECG are in the range of 1 µV to 100 mV, which have to be amplified using bioamplifiers for the acquisition and analyses (Nagel, 2000). Good bioamplifiers must fulfill the following criteria:

- Do not affect the physiological systems
- Do not distort the acquired signals
- Amplify the signals without noise and interference
- Free from electrical shock for the patient under testing

In three-electrode measurement systems, one electrode is used as a reference electrode and another two electrodes are used as measurement electrodes for biological potential. These factors, such as 50 or 60 Hz power line interference signals, harmonics, interference produced from the tissue or electrode interface and noise, influence the amplification process (Nagel, 2000). Figure 3.12 shows the schematic design of biopotential amplifier.

In the preamplifier stage, most of the noise and interference of the measured signals can be eliminated (Nagel, 2000). Few considerations need to be taken into account while designing the preamplifiers, such as electrode potential, noise, the connection between the amplifier and biological sources and electromagnetic interference. The high- and low-pass filters are used to eliminate the interference in the

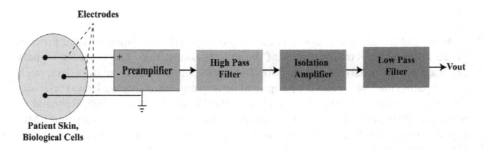

FIGURE 3.12   Biosignal amplifier.

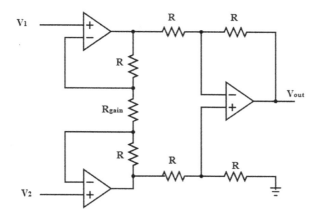

FIGURE 3.13 Instrumentation amplifier.

acquired signals, such as preamplifier offset potential, electrodes half-cell potential and to reduce the noise in a particular frequency range. The isolation amplifiers are used to protect the measurement devices and patients. The isolation amplifier is designed with one of the three different techniques, namely, the transformer isolation, opto isolation and capacitor isolation (Nagel, 2000).

The instrumentation amplifiers are used in the biological measurement systems for the amplification process. In medical instrumentation, the CMRR has to be more than 90 dB and the voltage gain must be greater than 80 dB for the biosignal amplification process. Instrumentation amplifiers are widely used for the amplification process because of inherent properties such as low DC offset, high gain, high CMRR, high accuracy and stability. Figure 3.13 shows the circuit diagram of a typical instrumentation amplifier.

3.10 FILTERS

In biological signals, filtering removes the signal components with undesired frequencies and keeping the signal components with desired frequencies (Palaniappan, 2011). In general, four types of filters, namely, low-pass filter, high-pass filter, band-pass filter and band-stop filter are available for various signal processing applications. Figure 3.14 show the pass band for the different filter used for filtering biological signals.

Low-pass filter blocks the higher frequency above the cut-off frequency and allows the low-frequency components below the cut-off frequency (Palaniappan, 2011). High-pass filter blocks the low frequency below the cut-off frequency and allows the higher-frequency components above the cut-off frequency. Band-pass filter allows all the frequency components between the lower and higher cut-off frequencies. Band-stop filter prevents all the frequency components between the lower and higher cut-off frequencies (Palaniappan, 2011).

The measurement of electrophysiological signals has the potential to offer an insight into several aspects of the physiological processes. These electrical signals recorded from the physiological system are well correlated with the mechanical processes associated

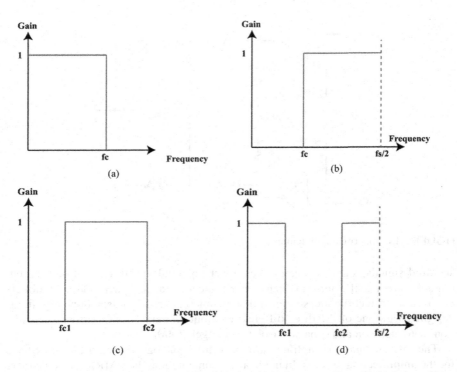

FIGURE 3.14   Different filters for biological signals (a) low-pass filter, (b) high-pass filter, (c) band-pass filter and (d) band-stop filter.

with the system. For example, the electrogastrograms acquired from the stomach region are well correlated with the physiological process of digestion as well as with various digestive pathologies (Alagumariappan et al., 2020a; Alagumariappan et al., 2016; Paramasivam et al., 2018; Rajagopal et al., 2018). The proper design of the instrumentation system for acquisition of biosignals plays the key role in the quality of the recorded biosignals. The design involves the proper selection of the electrode types, electrode systems, the ICs for instrumentation amplifiers and filter design (Alagumariappan & Krishnamurthy, 2018; Alagumariappan et al., 2020b). The design must be based on the characteristics of the biosignal under investigation and the skin-electrode model.

REFERENCES

Acharya, U. R., Ng, E. Y. K., Swapna, G., & Michelle, Y. S. L. (2011). Classification of normal, neuropathic, and myopathic electromyograph signals using nonlinear dynamics method. *Journal of Medical Imaging and Health Informatics*, *1*(4), 375–380.

Akay, M. (1996). *Detection and estimation methods for biomedical signals* (pp. 83–106). San Diego, CA: Academic Press.

Alagumariappan, P. & Krishnamurthy, K. (2018). An approach based on information theory for selection of systems for efficient recording of electrogastrograms. In: *Proceedings of the International Conference on Computing and Communication Systems* (pp. 463–471). Singapore: Springer.

Alagumariappan, P, Krishnamurthy, K, Kandiah, S, Cyril, E. & Venkatesan, R. (2020a) Diagnosis of type 2 diabetes using electrogastrograms: Extraction and genetic algorithm–based selection of informative features. *JMIR Biomedical Engineering*, *5*(1), e20932. doi: 10.2196/20932.

Alagumariappan, P., Krishnamurthy, K., & Jawahar, P. M. (2020b). Selection of surface electrodes for electrogastrography and analysis of normal and abnormal electrogastrograms using Hjorth information. *International Journal of Biomedical Engineering and Technology*, *32*(4), 317–330.

Alagumariappan, P., Rajagopal, A., & Krishnamurthy, K. (2016, October). Complexity analysis on normal and abnormal electrogastrograms using Tsallis entropy. In: *3rd International Electronic and Flipped Conference on Entropy and Its Applications*. Multidisciplinary Digital Publishing Institute. doi: 10.3390/ecea-3-A003.

Alberts, B., Bray, D., Lewis, J., Raff, M., Roberts, K., & Watson, J. D. (1994). *Molecular biology of the cell*. New York: Garland Publishing Inc.

Albulbul, A. (2016). Evaluating major electrode types for idle biological signal measurements for modern medical technology. *Bioengineering*, *3*(3), 20.

Bachoud-Lévi, A. C. (2017). From open to large-scale randomized cell transplantation trials in Huntington's disease: Lessons from the multicentric intracerebral grafting in Huntington's disease trial (MIG-HD) and previous pilot studies. *Progress in Brain Research*, *230*, 227–261, Elsevier.

Begg, R., Lai, D. T., & Palaniswami, M. (2007). *Computational intelligence in biomedical engineering*. CRC Press.

Bhoi, A. K., Sherpa, K. S., & Khandelwal, B. (2015). Multidimensional analytical study of heart sounds: A review. *International Journal Bioautomation*, *19*(3), 351–376.

Blottner, D., & Salanova, M. (2015). Neuromuscular System. In: *The neuromuscular system: From earth to space life science* (pp. 63–87). Cham: Springer.

Bronzino, J. D. (2000). *Biomedical engineering handbook* (Vol. 2). Berlin: Springer Science & Business Media.

Brown, Z. & Gupta, B. (2008). Biological signals and their measurement. *Update in Anaesthesia*, *24*(2), 164–169.

Bruce, E. N. (2001). *Biomedical signal processing and signal modeling* (pp. 335–336). New York: Wiley.

Carter, M. & Shieh, J. (2015). Electrophysiology. In: Guide to research techniques in neuroscience (pp. 89–115). Cambridge, MA: Academic Press.

Cattani, C., Badea, R., Chen, S., & Crisan, M. (2012). Biomedical signal processing and modeling complexity of living systems.

Chua, K. C., Chandran, V., Acharya, U. R., & Lim, C. M. (2010). Application of higher order statistics/spectra in biomedical signals: A review. *Medical Engineering and Physics*, *32*(7), 679–689.

Davis, P. J., & Cladis, F. P. (Eds.). (2016). *Smith's anesthesia for infants and children e-book*. Elsevier Health Sciences.

De Cooman, T., Goovaerts, G., Varon, C., Widjaja, D., & Van Huffel, S. (2014, September). Heart beat detection in multimodal data using signal recognition and beat location estimation. *In Proceedings* of the 41st Annual Computing in Cardiology (CinC2014) (pp. 257–260). Cambridge, MA: IEEE.

Deenen, J. C., Horlings, C. G., Verschuuren, J. J., Verbeek, A. L., & van Engelen, B. G. (2015). The epidemiology of neuromuscular disorders: a comprehensive overview of the literature. *Journal of Neuromuscular Diseases*, *2*(1), 73–85.

Duque, C. J. G., Muñoz, L. D., Mejía, J. G., & Trejos, E. D. (2014, September). Discrete wavelet transform and k-nn classification in EMG signals for diagnosis of neuromuscular disorders. In *2014 XIX symposium on image, signal processing and artificial vision*, Colombia (pp. 1–5). IEEE.

Enderle, J. & Bronzino, J. (2012). *Introduction to biomedical engineering*. Cambridge, MA: Academic Press.

Ghosh, S., Feng, M., Nguyen, H., & Li, J. (2014, September). Predicting heart beats using co-occurring constrained sequential patterns. *In Proceedings of the* Computing in Cardiology (pp. 265–268). Cambridge, MA: IEEE.

Gierałtowski, J. J., Ciuchciński, K., Grzegorczyk, I., Kośna, K., Soliński, M. & Podziemski, P. (2014). Heart rate variability discovery: Algorithm for detection of heart rate from noisy, multimodal recordings. *Computing in Cardiology*, 41, 253–256, IEEE.

Greenway, M. J., Andersen, P. M., Russ, C., Ennis, S., Cashman, S., Donaghy, C., ... & Morrison, K. E. (2006). ANG mutations segregate with familial and sporadic amyotrophic lateral sclerosis. *Nature Genetics*, *38*(4), 411–413.

Ha, S., Kim, C., Chi, Y. M., & Cauwenberghs, G. (2014). Low-power integrated circuit design for Wearable biopotential sensing. In: *Wearable sensors* (pp. 323–352). Cambridge, MA: Academic Press.

Hammond, C. (2014). *Cellular and molecular neurophysiology*. Cambridge, MA: Academic Press.

Hassan, M. F. U., Lai, D., & Bu, Y. (2019, October). Characterization of single lead continuous ECG recording with various dry electrodes. In: *Proceedings of the 2019 3rd International Conference on Computational Biology and Bioinformatics* (pp. 76–79). Japan: Naoya.

Hodgkin, A. L. & Huxley, A. F. (1952). A quantitative description of membrane current and its application to conduction and excitation in nerve. *The Journal of Physiology*, *117*(4), 500.

Holstege, G., Beers, C. M., & Subramanian, H. H. (2014). *The central nervous system control of respiration*. Amsterdam: Elsevier.

Horikawa, Y. (1998). Bifurcations in the decremental propagation of a spike train in the Hodgkin-Huxley model of low excitability. *Biological Cybernetics*, *79*(3), 251–261.

Humphrey, J. D. (2002). *Cardiovascular solid mechanics: Cells, tissues and organs*. New York: Springer-Verlag.

Humphrey, J. D. & McCulloch, A. D. (2003). The cardiovascular system: Anatomy, physiology and cell biology. In: Holzapfel, G. A. & Ogden, R. W. (Eds), *Biomechanics of soft tissue in cardiovascular systems* (pp. 1–14). Vienna: Springer.

Jamal, M. Z. (2012). Signal acquisition using surface EMG and circuit design considerations for robotic prosthesis. *Computational Intelligence in Electromyography Analysis-A Perspective on Current Applications and Future Challenges*, *18*, 427–448.

James, C. J. & Hesse, C. W. (2004). Independent component analysis for biomedical signals. *Physiological Measurement*, *26*(1), R15.

Jung, T. P., Makeig, S., Lee, T. W., McKeown, M. J., Brown, G., Bell, A. J., & Sejnowski, T. J. (2000, June). Independent component analysis of biomedical signals. In: Proce*edings of* Inter*national* Workshop on Independent Component Analysis and Signal Separation (pp. 633–644). Helsinki, Finland: Citeseer.

Kaniusas, E. (2012). Fundamentals of biosignals. In: *Biomedical signals and sensors I* (pp. 1–26). Berlin, Heidelberg: Springer.

Khandpur, R. S. (1987). *Handbook of biomedical instrumentation*. New York: McGraw-Hill Education.

Koslow, S. H. & Subramaniam, S. (2005). *Databasing the brain*. Hoboken, NJ: Wiley-Liss.

Kozlíková, K. (2010, January). Biological signals in medical diagnostics. In: AIP Conference Proceedings (Vol. 1204, No. 1, pp. 147–150). American Institute of Physics. doi: 10.1063/1.3295626.

Lessard, C. S. (2005). Signal processing of random physiological signals. *Synthesis Lectures on Biomedical Engineering*, *1*(1), 1–232.

Macy, A. (2015). *The handbook of human physiological recording*. Available online: https://alanmacy.com/book/the-handbook-of-human-physiological-recording/

Majdalawieh, O., Gu, J., Bai, T., & Cheng, G. (2003, August). Biomedical signal process-ing and rehabilitation engineering: A review. In: *2003 IEEE Pacific Rim Conference on Communications Computers and Signal Processing (PACRIM 2003)*, Victoria, BC, Canada, (Vol. 2, pp. 1004–1007). IEEE.

Manahan-Vaughan, D. (2018). *Handbook of in vivo neural plasticity techniques: A systems neuroscience approach to the neural basis of memory and cognition*. Cambridge, MA: Academic Press.

Marieb, E. N. & Hoehn, K. (2007). *Human anatomy and physiology*. London: Pearson Education.

Maršálek, P. (2000). Coincidence detection in the Hodgkin–Huxley equations. *Biosystems, 58*(1–3), 83–91.

Marston, H. R., Hadley, R., Banks, D., & Duro, M. D. C. M. (2019, September). Mobile self-monitoring ECG devices to diagnose arrhythmia that coincide with palpitations: A scop-ing review. Healthcare (Basel), 7(3), 96, Multidisciplinary Digital Publishing Institute.

Martini, F. (2006). *Anatomy and physiology'2007 Ed.* Philippines: Rex Bookstore, Inc.

McCormick, D. A. (2014). Membrane potential and ction potential. In: Roberts, J. L. & Byme, J. H. (Eds), *From molecules to networks* (pp. 351–376). Cambridge, MA: Academic Press.

Milnor, W. R. (1990). *Cardiovascular physiology*. Oxford: Oxford University Press.

Mitov, I. P. (1998). A method for assessment and processing of biomedical signals containing trend and periodic components. *Medical Engineering and Physics, 20*(9), 660–668.

Nagel, J. H. (2000). *Biopotential amplifiers: Heart rate variability*. Boca Raton, FL: CRC Press LLC.

Naït-Ali, A. (Ed.). (2009). *Advanced biosignal processing*. Berlin, Germany: Springer Science & Business Media.

Opie, L. H. (1998). The heart-physiology from cell to circulation. *Acta Cardiologica, 53*, 127–127.

Palaniappan, R. (2011). *Biological signal analysis*. London: BookBoon.

Paramasivam, A., Kamalanand, K., Emmanuel, C., Mahadevan, B., Sundravadivelu, K., Raman, J., & Jawahar, P. M. (2018, March). Influence of electrode surface area on the fractal dimensions of electrogastrograms and fractal analysis of normal and abnormal digestion process. *In 2018 International Conference on Recent Trends in Electrical, Control and Communication (RTECC)* (pp. 245–250). IEEE, Chennai, India.

Penzel, T., Kemp, B., Klosch, G., Schlogl, A., Hasan, J., Varri, A., & Korhonen, I. (2001). Acquisition of biomedical signals databases. *IEEE Engineering in Medicine and Biology Magazine, 20*(3), 25–32.

Plieth, W. (2008). *Electrochemistry for materials science*. Amsterdam: Elsevier.

Preejith, S. P., Dhinesh, R., Joseph, J. & Sivaprakasam, M. (2016). Wearable ECG platform for continuous cardiac monitoring. *In 2016 38th Annual International Conference of the IEEE Engineering in Medicine and Biology Society (EMBC)* (pp. 623–626). IEEE, Orlando.

Rajagopal, A., Alagumariappan, P., & Krishnamurthy, K. (2018). Development of an auto-mated decision support system for diagnosis of digestive disorders using electrogastro-grams: An approach based on empirical mode decomposition and K-means algorithm. In: Sarraf, J., Pattnaik, P. K., & Swetapadma, A. (Eds), *Expert system techniques in biomedical science practice* (pp. 97–119). Pennsylvania: IGI Global.

Reed, S. M., Bayly, W. M., & Sellon, D. C. (2017). *Equine internal medicine-e-book*. Elsevier Health Sciences.

Robergs, R. A., & Roberts, S. (1996). *Exercise physiology: Exercise, performance, and clini-cal applications* (pp. 546–563). St. Louis: Mosby.

Rowland, L. P., & Shneider, N. A. (2001). Amyotrophic lateral sclerosis. *New England Journal of Medicine, 344*(22), 1688–1700.

Rubin, D. I. (2019). Needle electromyography: Basic concepts. In: Levin, K. and Chauvel, P. (Eds.) Handbook of clinical neurology (Vol. 160, pp. 243–256). Amsterdam: Elsevier.

Rudroff, T. (2008). *Kinesiological fine wire EMG: A practical introduction to fine wire EMG applications.* Arisona: Noraxon.

Sadikoglu, F., Kavalcioglu, C., & Dagman, B. (2017). Electromyogram (EMG) signal detection, classification of EMG signals and diagnosis of neuropathy muscle disease. *Procedia Computer Science, 120,* 422–429.

Saladin, K. S. (2004). *Anatomy and physiology: The unity of form and function.* New York: McGraw-Hill Education.

Selvan, V. A. (2011). Single-fiber EMG: A review. *Annals of Indian Academy of Neurology, 14*(1), 64.

Semmlow, J. L. & Griffel, B. (2014). *Biosignal and medical image processing.* Boca Raton, FL: CRC Press.

Serhani, M. A., El Kassabi, T., Ismail, H., & Nujum Navaz, A. (2020). ECG monitoring systems: Review, architecture, processes, and key challenges. *Sensors, 20*(6), 1796.

Singh, M. (2014). *Introduction to biomedical instrumentation.* New Delhi: PHI Learning Pvt. Ltd.

Smelser, N. J., & Baltes, P. B. (Eds.). (2001). *International encyclopedia of the social & behavioral sciences* (Vol. 11). Amsterdam: Elsevier.

Stålberg, E. (1980). Macro EMG, a new recording technique. *Journal of Neurology, Neurosurgery and Psychiatry, 43*(6), 475–482.

Subasi, A. (2019). *Practical guide for biomedical signals analysis using machine learning techniques: A MATLAB based approach.* Cambridge, MA: Academic Press.

Thakor, N. V. (2015). Biopotentials and electrophysiology measurements. In: Webster, J. G. and Eren, H. (Eds.) *Telehealth and mobile health* (pp. 595–614). Boca Raton, FL: CRC Press.

Togawa, T., Tamura, T., & Oberg, P. A. (1997). *Biomedical transducers and instruments.* Boca Raton, FL: CRC Press.

Tortora, G. J. & Derrickson, B. H. (2018). *Principles of anatomy and physiology.* Hoboken, NJ: John Wiley & Sons.

Van Drongelen, W. (2010). *Signal processing for neuroscientists, a companion volume: Advanced topics, nonlinear techniques and multi-channel analysis.* Amsterdam: Elsevier.

Van Drongelen, W. (2018). *Signal processing for neuroscientists.* Cambridge, MA: Academic Press.

Webster, J. G. (1984). Reducing motion artifacts and interference in biopotential recording. *IEEE Transactions on Biomedical Engineering, 31*(12), 823–826.

White, J. A. (2002). Action potential. In: Ramachandran, V. (Ed.) Encyclopedia of *the human brain* (pp. 1–12). Cambridge, MA: Academic Press.

World Health Organization. Cardiovascular diseases. Available online: https://www.who.int/health-topics/cardiovascular-diseases/#tab=tab_1. (accessed on 13 August 2020).

Yazıcıoğlu, R. F., Van Hoof, C., & Puers, R. (2009). Introduction to biopotential acquisition. In: Hoof, C. V., Yazicioglu, R. F., & Puers (Eds.) *Biopotential readout circuits for portable acquisition systems* (pp. 5–19). Dordrecht: Springer.

Zhang, J. X. & Hoshino, K. (2018). *Molecular sensors and nanodevices: Principles, designs and applications in biomedical engineering.* Cambridge, MA: Academic Press.

4 Electromyograms

4.1 NEUROMUSCULAR SYSTEM AND ELECTROMYOGRAPHY

In the complex human physiological system, the muscular and nervous system jointly form the neuromuscular system that produces the contraction of muscles (Reed et al., 2017; Smelser & Baltes, 2001; Davis et al., 2016). The human neuromuscular system is a complex electromechanical system associated with many subsystems and interconnected complex variables (Robergs & Roberts, 1996; Begg et al., 2007). The motor neurons, Schwann cells, neuromuscular junction and several types of muscle fibers are the major components of the neuromuscular system (Smelser & Baltes, 2001). Further, the neuromuscular junction is categorized into three types, namely, synaptic, postsynaptic and presynaptic space.

Essential functions of the neuromuscular system are the movement of the body, the respiratory movements (inhalation and exhalation) and posture control (Smelser & Baltes, 2001; Robergs & Roberts, 1996). The neuromuscular junction releases the neurotransmitter acetylcholine molecules when electrical potentials are generated from motor neurons and propagated along with the motor axon (Reed et al., 2017). The muscle membrane potentials are generated by the diffusion of neurotransmitter acetylcholine molecules with receptor acetylcholine molecules in the muscle fiber. This muscle membrane potential stimulates a calcium ion discharge that produces the muscle movements (Robergs & Roberts, 1996; Reed et al., 2017).

The disorders that affect the muscular system as well as the nervous system (neuromuscular system) are referred to as the neuromuscular disorders (Blottner & Salanova, 2015). Several disorders such as myopathies (problems associated with muscles), neuropathies (problems associated with the nerves), amyotrophic lateral sclerosis (ALS), multiple sclerosis and myasthenia gravis affect the human neuromuscular system (Blottner & Salanova, 2015; Rowland & Shneider, 2001; Greenway et al., 2006; Garmirian et al., 2009). Every year, the number of individuals suffering from neuromuscular disorders increases exponentially, which results in increased mortality rate and decreased life span. Myopathy and ALS are the two main groups of neuromuscular disorders (Artameeyanant et al., 2014; Preston & Shapiro, 2012). The disorders associated with muscle and its functions are referred to as myopathy, and the common symptoms of myopathies are muscle weakness, cramp, fatigue, myotonia, muscle atrophy and spasm (Artameeyanant et al., 2014; Preston & Shapiro, 2012). Further, myopathies are classified into two types, namely, inherited and acquired myopathies (Artameeyanant et al., 2014). Several myopathies such as inflammatory myopathies, toxic myopathies, congenital myopathies, endocrine myopathies and metabolic myopathies are associated with muscle pathologies (Preston & Shapiro, 2012). The disorders associated with the lower and upper motor neurons are referred to as neuropathy. ALS is typically a neurodegenerative motor neuron disorder which leads to death from respiratory failure (Greenway et al., 2006; Rowland & Shneider, 2001).

Further, ALS is classified into two types, namely, sporadic ALS (Greenway et al., 2006) and hereditary ALS (Couratier et al., 2016).

The electrical signals from the neuromuscular system are recorded using the standard electrodiagnostic technique known as "electromyography," which is a major diagnostic tool for the confirmation of the neuromuscular disorders (Sadikoglu et al., 2017; Duque et al., 2014; Greenway et al., 2006; Rowland & Shneider, 2001; Blottner & Salanova, 2015; Ambikapathy & Krishnamurthy, 2018; Ambikapathy et al., 2018; Bakiya & Kamalanand, 2018; Bakiya et al., 2020). The EMG signals are acquired either using surface electrodes which are non-invasive electrodes and well acceptable to the patients (Smith & Hargrove, 2013; Greenway et al., 2006; Sadikoglu et al., 2017; Day, 2002) or needle electrodes which are invasive (electrodes inserted into the muscle fiber) (Smith & Hargrove, 2013; Greenway et al., 2006; Rowland & Shneider, 2001).

Further, needle electrodes are classified into four types, namely, concentric needle, monopolar needle, single fiber needle and macro needle electrodes. The measurement of EMG signals using invasive electrodes causes discomfort to the patients (Jost et al., 1994), but the measurement of EMG signals using non-invasive electrodes are comfortable and does not cause any discomfort or pain to the patients (Smith & Hargrove, 2013). In recent years, several researchers have focused on selecting proper electrodes for electrophysiological measurements using information analysis methods (Ghuman & DeBrunner, 2014; Smith & Hargrove, 2013). Figure 4.1 shows the typical electromyograms acquired using various types of electrodes.

4.2   EFFECT OF ELECTRODE TYPES ON THE ACQUIRED EMG SIGNALS

The information analysis methods such as entropy measures are utilized for the analysis of the electrophysiological measurement systems. Entropy (Singh, 2013; Alagumariappan et al., 2017) is a quantification of the degree of disorder associated with a system and hence, it is a measure of the information content associated with the system under investigation. The general thermodynamic entropy is expressed as:

$$\text{Entropy} = k \log D \tag{4.1}$$

where D is the measure of uncertainty and k is the Boltzmann constant. Shannon expressed entropy as an amount of information gained and, hence, as a measure of uncertainty (Singh, 2014). The Shannon entropy is expressed as:

$$\text{Shannon entropy} = \sum_{i=1}^{n} P(x_i) \log_a \frac{1}{p(x_i)} \tag{4.2}$$

where $p(x_i)$ is the probability of incidence of x_i. Related to Shannon entropy, the simplification of Shannon entropy is defined as the Tsallis entropy with order of α (Singh 2014),

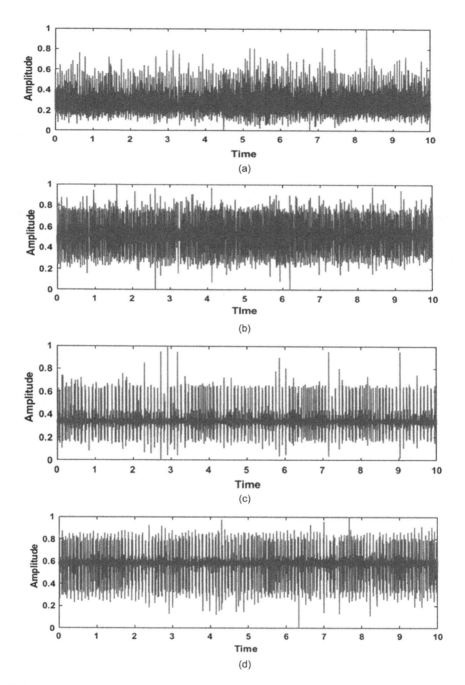

FIGURE 4.1   Typical EMG signals acquired using various types of electrodes: (a) monopolar needle electrodes, (b) concentric needle electrodes, (c) surface electrodes and (d) fine-wire electrodes.

$$H_{\text{Tsallis}} = \frac{1}{1-\alpha}\left(1 - \sum_{i-1}^{n} p(x_i)^{\alpha}\right) \qquad (4.3)$$

Figure 4.2 shows the variation of Tsallis entropy values of electromyogram signals acquired using surface, fine wire, monopolar needle and concentric needle electrodes with respect to different α values. It is observed in all the cases that the Tsallis entropy of EMG signals decreases exponentially with respect to α values ranging from 0.2 to 5. The Tsallis entropy is higher in the case of electromyograms recorded with the help of concentric needle electrodes compared to that of the electromyograms recorded with surface electrodes, fine-wire electrodes and monopolar needle electrodes. Moreover, Tsallis entropy is higher in the case of EMG signals recorded using needle electrodes compared to that of the signals recorded using surface and fine-wire electrodes. Hence, the information of the EMG signals recorded using needle electrodes is higher than that of the EMG signals acquired using fine-wire and surface electrodes.

Further, the Rényi entropy with order of α is given by Singh (2014),

$$H_{\text{Renyi}} = \frac{1}{1-\alpha}\ln\sum_{i-1}^{n} p(x_i)^{\alpha} \qquad (4.4)$$

The changes in the Rényi entropy values of EMG signals acquired using different types of electrodes (surface, fine wire, monopolar needle and concentric needle) are shown as a function of α ranging from 0.1 to 0.9, as shown in Figure 4.3. The mean Rényi entropy value is higher in the cases of EMG signals measured using concentric needle electrodes compared to that of EMG signals acquired using fine-wire, monopolar needle and surface electrodes. Further, the average Rényi entropy of the EMG

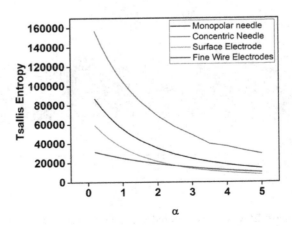

FIGURE 4.2 The variation of Tsallis entropy values of electromyogram signals acquired using different types of electrodes shown as a function of α.

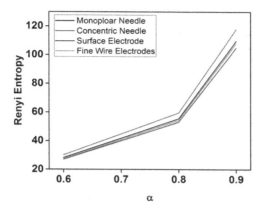

FIGURE 4.3 Variation of Rényi entropy values of EMG signals acquired using different types of electrodes shown as a function of α.

signals acquired using needle electrodes is found to be higher than the average Rényi entropy of the EMG signals acquired using fine-wire and surface electrodes.

Spectral entropy is the frequency-domain entropy which measures the signal power irregularities over measured frequencies. The spectral entropy is expressed as (Zawawi et al., 2013):

$$\text{SEN} = \frac{-\sum p_k \log p_k}{\log(N)} \tag{4.5}$$

where p_k is the normalized spectral power of the frequency and N is the number of frequencies. Figure 4.4 shows the variation of spectral entropy of electromyograms

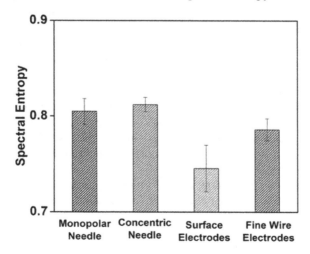

FIGURE 4.4 The difference in spectral entropy values of EMG signals measured using different types of electrodes.

acquired using fine-wire, monopolar needle, concentric needle and surface electrodes. It is observed that the values of spectral entropy are higher in the case of electromyograms measured using concentric needle electrodes compared to that of spectral entropy of electromyograms acquired using fine-wire electrodes, monopolar needle electrodes and surface electrodes. Hence, the EMG signals acquired using invasive electrodes exhibit higher entropy compared to those acquired using non-invasive electrodes.

In the measurement of EMG signals, there are two categories of contractions, namely, isotonic contraction and isometric contraction (Nazmi et al., 2016). Isotonic contraction is often used for athletic goals, whereas isometric contractions are utilized for physical rehabilitation. The isotonic contraction is a muscle contraction that produces force with respect to the resistance in which the length of the muscle changes. Further, isotonic contraction is classified into two types, namely, the concentric contraction and eccentric contraction (Nazmi et al., 2016). The combination of concentric and eccentric contraction creates a dynamic contraction (e.g., the joint movement during dumbbell bicep curl exercise). Concentric contraction permits the muscle to reduce unstable energy but rigidity remains constant during the contraction. In eccentric contraction, the length of the contraction is longer, disturbing the muscles to stretch in response to a greater opposing force. Isometric contraction is the contraction that creates no change in muscle length but the energy and tension remain unstable (Nazmi et al., 2016).

4.3   TYPES OF MUSCLES

The human muscular system is a complex system comprising skeletal muscle, cardiac muscle and smooth muscle, with each type of muscle having specific characteristics (Saladin & McFarland, 2008; Kamen & Gabriel, 2010). The muscular system permits the body movement, blood circulation and posture of the human body (Kamen & Gabriel, 2010). Figure 4.5 shows the different types of muscles in the physiological system.

Smooth muscles are non-striated muscles, and humans cannot control the activities of the smooth muscles (Bohr, 1973). Smooth muscle is spindle shaped and comprises cells ranging from 30 to 200 µm with central single nucleus. There are two types of smooth muscles, namely, multiunit smooth muscle and visceral smooth muscle (Bohr, 1973). The visceral smooth muscle is commonly known as unitary smooth muscle and produce slow and steady contraction such as the intestinal motility which allow food transport from the proximal to the distal loop. The multi-unit smooth muscles are observed in the lungs, arteries, internal eye muscle, erector pili muscle etc. (Bohr, 1973).

Cardiac muscles are also known as heart muscles and are controlled by the autonomic nervous system (Olivetti et al., 1996). There are three categories of cardiac muscles, namely, atrial muscle, ventricular muscle and excitatory muscles. The atria and ventricles of the heart are formed of the cardiac smooth muscles. Though the tissues of the atria and ventricles are similar, some differences exist. The myocardium in the ventricles is thick whereas myocardium in the atria is thinner (Olivetti et al., 1996).

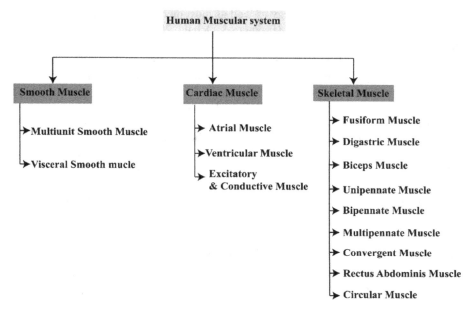

FIGURE 4.5 Types of muscles in the human muscular system.

The skeletal muscle is otherwise referred to as striated muscle and is associated with the somatic nervous system (Lieber, 2002; Frontera & Ochala, 2015). The human skeleton has approximately 640 skeletal muscles (320 pairs) which are categorized into different groups: the head muscles, neck muscles, muscles of the torso and the muscles of the upper and lower extremities. (MacIntosh et al., 2006). The shape of skeletal muscles is categorized into four different groups as parallel, convergent, circular and pennate (MacIntosh et al., 2006). Most skeletal muscles are parallel muscles and have different shapes, flat bands, spindle shaped and belly shaped. The parallel muscles are characterized by fascicles running parallel to each other, and are also classified into two types based on their shapes, namely, fusiform and non-fusiform muscles. The fusiform muscles are spindle-shaped structures whereas non-fusiform muscles are rectangular shaped with constant diameter. Biceps brachial muscle is the most common parallel muscle in skeletal muscles.

Convergent muscles are otherwise known as triangular muscles, and the muscle fibers spread over a broad area but all muscle fibers converge into a common narrow tendon. The pectoralis major is a common convergent muscle which found in the chest.

In pennate muscles, the tendons run throughout the length of the muscles. Further, they are categorized into three types: unipennate (all the fascicles are on the same side of the tendon), bipennate (fascicles lie on either side of the tendon) and multipennate (central tendon branches within a pennate muscle). The rectus femoris is a pennate muscle found in thigh.

Circular muscles or sphincter muscles are organized concentrically around an opening or recess. The orbicularis oris is an example of circular muscles which is found around the mouth.

Skeletal muscles produce energy as well as movement of muscles, and their morphological unit is the muscle fiber (Frontera & Ochala, 2015). The muscle myofiber is otherwise known as muscle cell and is cylindrical in shape with 10–100 μm diameter and 1.5 mm to 30 cm length (Wang et al., 2008; MacIntosh et al., 2006). The myofiber bundles are joined by muscle tissue which are responsible for force and movement (MacIntosh et al., 2006).

The skeleton muscles are attached to the bones by tendons (bundle of collagen fibers) in an area called insertion zone (Thorpe et al., 2013), which is usually distant form the distal portion of the muscle because the movement of muscle occurs at the joint (Lu & Jiang, 2005; MacIntosh et al., 2006; Frontera & Ochala, 2015). Tendons are highly capable of bearing tension and strong band tissue connection (links one end of the skeletal muscle to joint or bone). The functions of the each muscle in the muscular system are controlled by the nervous system (Lu & Jiang, 2005).

4.4   CHARACTERISTICS OF EMG SIGNALS

Raw EMG signals have peak-to-peak amplitude of 0–5 mv with the band frequency range of 0–1,000 Hz. Thus, the significant information pattern lies in range 20–500 Hz. The dominant frequency range of EMG signals is between 50 and 150 Hz (Fattah et al., 2012; Ghapanchizadeh et al., 2017). Generally, the amplitude of the raw EMG signals is low with addition of noise components due to the impedance of the body's skin. Several factors generate noises during recording such as fat between skin and muscle, blood flow in minuscule vessels, electromagnetic radiation sources, electrical wires, fluorescent lamp, subcutaneous tissue layer, spread of innervation zone, crosstalk, electrode size and electrode position (Soderberg & Cook, 1984; Van Boxtel, 2001). Figure 4.6 presents the characteristic features of normal, myopathy and ALS electromyograms.

The interpretation of the EMG depends on the observed amplitudes and action potentials in each patient. In normal EMG signals, the peak amplitude is expected to be 400 μv with a pulse duration of 15 ms; in case of myopathy, it is expected to be around 250 μv with a pulse duration of 20 ms; and in case of ALS EMG, it is increases to 1,000 μv with a pulse duration of 10 ms.

In terms of firing rates, myopathic and neuropathic cases have higher firing rates when compared to the normal individuals in that age group (Artameeyanant et al., 2016; Nikolic, 2001). Generally, neurological experts analyze EMG signals as a time series by extracting specific parts of the signals and classifying them. The time series can also be transformed into the frequency domain. Such transformation describes the peak frequency and the frequency range of the EMG signals, but it is difficult to differentiate the myopathy and ALS EMG signals using only the values of peak frequency. Figure 4.7 presents the frequency spectrum of normal, myopathy and ALS electromyogram recordings. It shows that the peak frequency of normal, myopathy and ALS EMG signals are 77.5, 95.04 and 136.5 Hz, respectively. Though there are variations in the peak frequencies, the discrimination of normal and abnormal EMG signals is difficult using frequency spectrum analyses without automation of the diagnosis process.

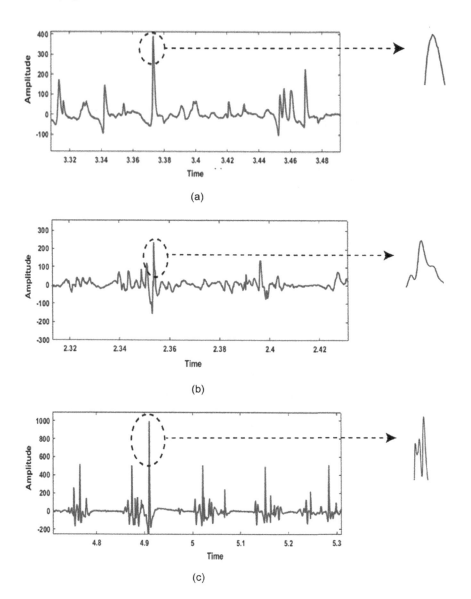

FIGURE 4.6 Characteristic of EMG signals: (a) normal, (b) myopathy and (c) ALS.

4.5 ELECTRODE PLACEMENT

Needle electrodes efficiently record the EMG signals of normal and abnormal individuals (Menkes & Pierce, 2019; Geiringer, 1999). Several types of needle electrodes are available based on the shapes and types in the clinical environment for EMG measurement system. Generally, needle electrodes are categorized into four types,

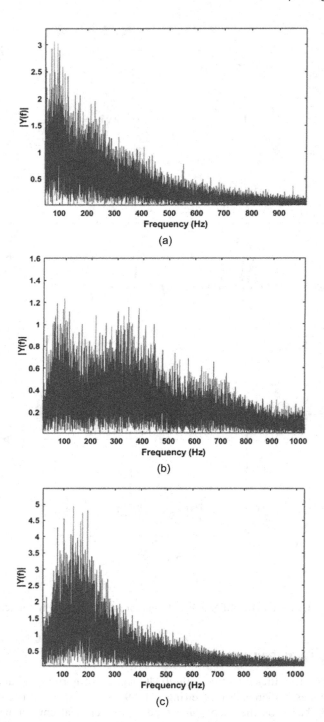

FIGURE 4.7   Frequency spectrum of EMG signals: (a) normal, (b) myopathy and (c) ALS.

namely, concentric needle electrodes, monopolar needle electrodes, single fiber and macro needle electrodes. Needle electrodes must be stable and sterile. All electrodes must undergo the low-power microscope test to determine the quality of electrodes. Sometimes, the disposable needle electrodes may have a very thin and poorly conducting film on the surface of the needle (Menkes & Pierce, 2019; Leis & Trapani, 2000). This conducting film increases the impedance that leads to low voltage, irregular and noisy signals. Mostly, researchers prefer concentric needle electrodes instead of other electrodes because concentric needle electrodes do not require a surface reference, the signal is crisper and analysis may be conducted more quickly (Menkes & Pierce, 2019; Boon et al., 2011). The muscle should be palpated by the examiner's thumb and index finger before inserting the needle into the muscle. This examination makes the skin taut to reduce bleeding. Generally, needles are available up to 120 mm in length and the length varies depending on the circumstances (Menkes & Pierce, 2019; Boon et al., 2008). The localization of the needle electrode insertion into upper and lower body muscles (Menkes & Pierce, 2019; Boon et al., 2008, 2011; Geiringer, 1999) is shown in Figure 4.8.

According to SENIAM (surface EMG for a non-invasive assessment of muscle), the standard electrode shape, electrode size, inter-electrode distance and electrode placement are recommended for the acquisition of EMG signals in the physiological system (Hermens et al., 1999). In the EMG measurement system, electrodes should be circular or square-shaped with a diameter of 10 mm. The conductive area of the surface electrodes varies from 1 mm^2 to a few mm^2. For bipolar electrodes, the inter-electrode distance (distance between two electrodes) is 20 mm (inter-electrode should not exceed one-fourth of the muscle fiber length) (Hermens et al., 2000). Electrodes materials such as Ag/AgCl, AgCl and Ag are used to acquire EMG signals based on the electrode skin impedance, and among them the Ag/AgCl electrodes (either non-gelled or electrode gel) are commonly used in the clinical environment (Merletti, 2000; Merletti et al., 2001; Soderberg & Cook, 1984). Figure 4.9a and b show the surface electrode placement to record EMG signals for various muscles in humans (Hermens et al., 1999, 2000; Merletti, 2000; Soderberg & Cook, 1984; De Luca, 1997).

4.6 COMPUTATIONAL ANALYSIS ON NORMAL AND ABNORMAL EMG SIGNALS

Figure 4.10 shows the typical normal, myopathy and ALS EMG signals as measured form biceps brachial muscle region over 11.8 s. The sampling rate of these acquired EMG signals is 23438.75 Hz (Nikolic, 2001). Further, the two-dimensional time-frequency images are obtained using transformation techniques such as spectrogram, Stockwell transform and Wigner–Ville transform techniques.

4.6.1 Spectrogram

A spectrogram is a time-frequency representation of a one-dimensional signal, which is an extension of the fast-Fourier transform. In general, the spectrograms provide spectral features in time domain and the distribution of energy in time-frequency domain.

The spectrograms can be expressed as reported by Zawawi et al. (2013),

$$S(t,f) = | \int_{-\infty}^{\infty} x(\tau)w(\tau - t)e^{-j2\pi \, \text{to} \, \tau} dt |^2 \qquad (4.6)$$

where $x(\tau)$ is the input signal and $w(t)$ is the window function as a function of time. Figure 4.11 shows the typical spectrograms of normal, myopathy and ALS electromyograms.

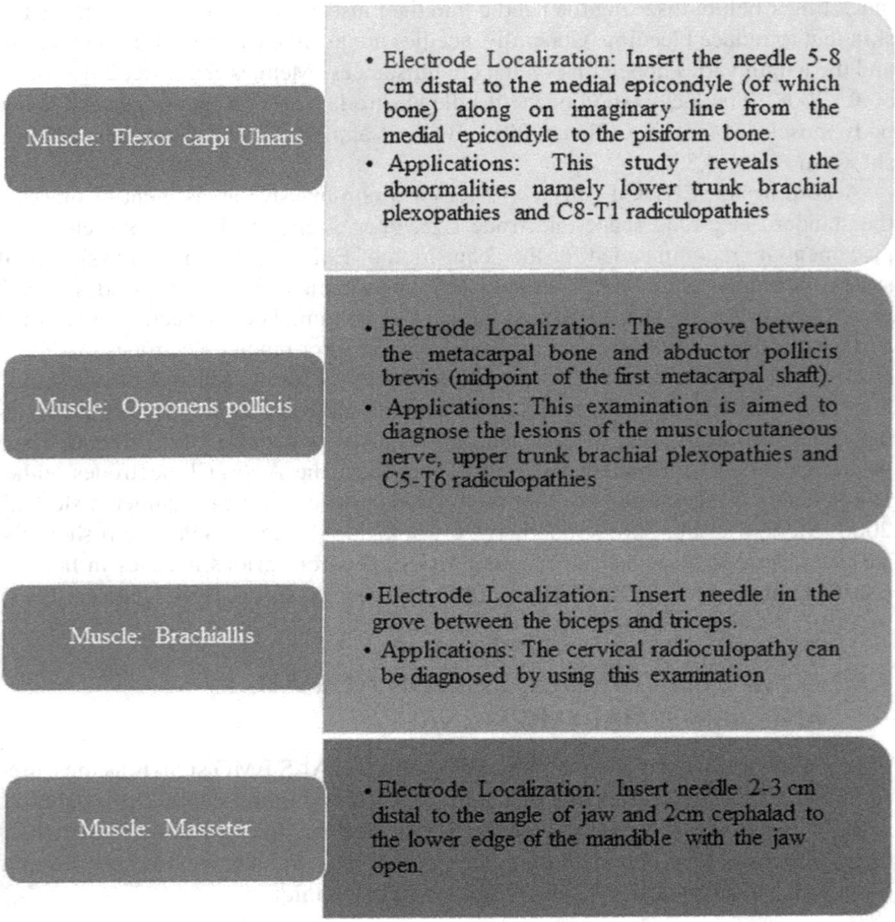

Muscle: Flexor carpi Ulnaris

- Electrode Localization: Insert the needle 5-8 cm distal to the medial epicondyle (of which bone) along on imaginary line from the medial epicondyle to the pisiform bone.
- Applications: This study reveals the abnormalities namely lower trunk brachial plexopathies and C8-T1 radiculopathies

Muscle: Opponens pollicis

- Electrode Localization: The groove between the metacarpal bone and abductor pollicis brevis (midpoint of the first metacarpal shaft).
- Applications: This examination is aimed to diagnose the lesions of the musculocutaneous nerve, upper trunk brachial plexopathies and C5-T6 radiculopathies

Muscle: Brachiallis

- Electrode Localization: Insert needle in the grove between the biceps and triceps.
- Applications: The cervical radioculopathy can be diagnosed by using this examination

Muscle: Masseter

- Electrode Localization: Insert needle 2-3 cm distal to the angle of jaw and 2cm cephalad to the lower edge of the mandible with the jaw open.

(a)

FIGURE 4.8 Localization of needle electrodes into the (a) upper body muscles and (b) lower body muscles.

(*Continued*)

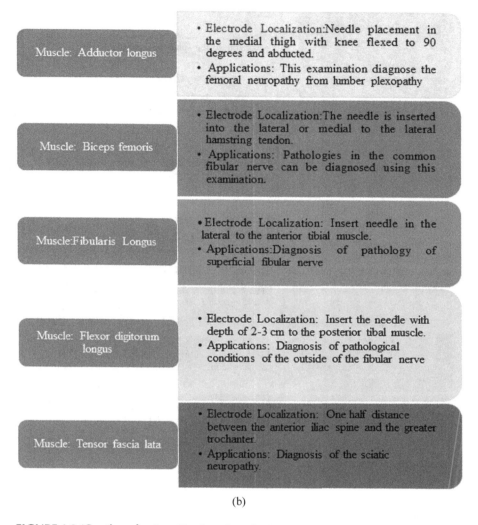

(b)

FIGURE 4.8 (*Continued*) Localization of needle electrodes into the (a) upper body muscles and (b) lower body muscles.

4.6.2 STATISTICAL ANALYSIS OF EMG SIGNALS

Statistical analysis is an important tool for the analysis of biosignals that provides the crucial mathematical interpretation of the acquired signals from the physiological system (Tenan et al., 2017). Statistical analysis is primarily performed to transform the huge set of observations into interpretable forms. The statistical measures such as, mean, median, variance, standard deviation, Hjorth parameters, fractal dimensions, skewness, kurtosis and Hurst exponent are used for the analysis of electromyogram signals.

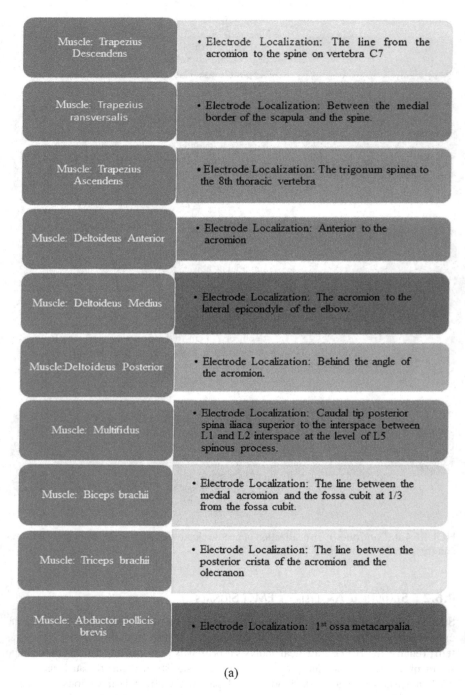

Muscle: Trapezius Descendens	• Electrode Localization: The line from the acromion to the spine on vertebra C7
Muscle: Trapezius ransversalis	• Electrode Localization: Between the medial border of the scapula and the spine.
Muscle: Trapezius Ascendens	• Electrode Localization: The trigonum spinea to the 8th thoracic vertebra
Muscle: Deltoideus Anterior	• Electrode Localization: Anterior to the acromion
Muscle: Deltoideus Medius	• Electrode Localization: The acromion to the lateral epicondyle of the elbow.
Muscle:Deltoideus Posterior	• Electrode Localization: Behind the angle of the acromion.
Muscle: Multifidus	• Electrode Localization: Caudal tip posterior spina iliaca superior to the interspace between L1 and L2 interspace at the level of L5 spinous process.
Muscle: Biceps brachii	• Electrode Localization: The line between the medial acromion and the fossa cubit at 1/3 from the fossa cubit.
Muscle: Triceps brachii	• Electrode Localization: The line between the posterior crista of the acromion and the olecranon
Muscle: Abductor pollicis brevis	• Electrode Localization: 1st ossa metacarpalia.

(a)

FIGURE 4.9 Localization of surface electrodes on (a) upper body muscles and (b) lower body muscles.

(Continued)

Muscle: Gluteus Maximus	• Electrode Localization: Between the sacral vertebrae and the greater trochanter.
Muscle: Tensor Fasciae Latae	• Electrode Localization: The anterior spina iliaca superior to the lateral femoral condyle in the proximal 1/6.
Muscle: Vastus medialis	• Electrode Localization: The anterior spina iliaca superior and the joint space in front of the anterior border of the medial ligament.
Muscle: Vastus lateralis	• Electrode Localization: The anterior spina iliaca superior to the lateral side of the patella.
Muscle: Biceps femoris	• Electrode Localization: Between the ischial tuberosity and the lateral epicondyle of the tibia
Muscle: Semitendinosus	• Electrode Localization: Between the ischial tuberosity and the medial epicondyle of the tibia..
Muscle: Tibialis anterior	• Electrode Localization: Between the tip of the fibula and the tip of the medial malleolus
Muscle: Peroneus longus	• Electrode Localization: Between the tip of the head of the fibula to the tip of the lateral malleolus.

(b)

FIGURE 4.9 (Continued) Localization of surface electrodes on (a) upper body muscles and (b) lower body muscles.

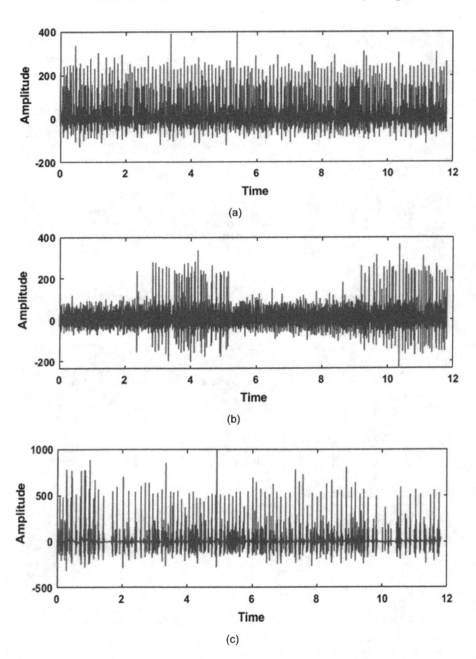

FIGURE 4.10 Typical EMG signals: (a) normal, (b) myopathy and (c) ALS.

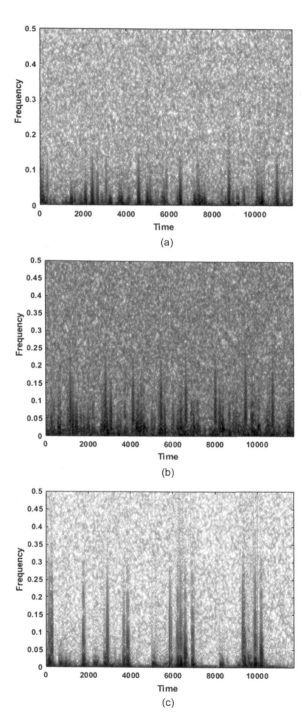

FIGURE 4.11 Spectrogram images of EMG signals: (a) normal, (b) myopathy and (c) ALS.

Descriptive statistics is commonly used for describing and analyzing large amounts of data (Gandhi & Sarkar, 2016). The statistical parameters such as mean, median, standard deviation, kurtosis and skewness are essential and highly useful in the preliminary characterization of normal and abnormal biosignals. Figure 4.12 shows the mean of normal, myopathy and ALS electromyograms. It is observed that the mean of myopathy EMG signal is higher compared with the mean of normal and ALS electromyograms. Figure 4.13 shows the median of normal, myopathy and ALS electromyograms. It is seen that the median of normal signal is higher when compared to the mean of myopathy and ALS electromyograms. Figure 4.14 shows the standard deviation of normal, myopathy and ALS EMG signals. It is seen that the standard deviation of normal signal is higher compared to the standard deviation of myopathy and ALS EMG signals.

Figure 4.15 shows the variation of skewness value of ALS, normal and myopathy electromyograms. It is seen that the skewness of ALS EMG signals is found to be higher when compared with the skewness of normal and myopathy EMG signals. In all cases, the degree of skewness is negatives (SK < 0) which indicates excess high values in the distribution. Figure 4.16 shows the variation of kurtosis of normal,

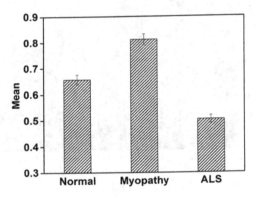

FIGURE 4.12   Variation in mean values of normal, myopathy and ALS EMG signals.

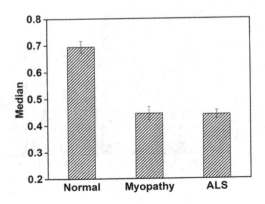

FIGURE 4.13   Median of normal, myopathy and ALS EMG signals.

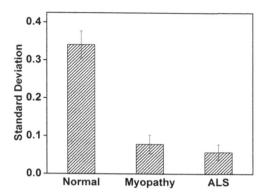

FIGURE 4.14 Standard deviation of normal, myopathy and ALS EMG signals.

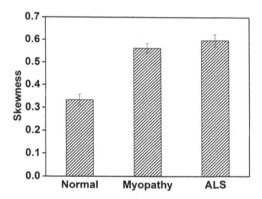

FIGURE 4.15 Variation of skewness values of normal, myopathy and ALS EMG signals.

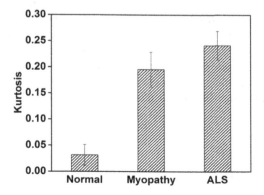

FIGURE 4.16 variation of kurtosis of normal, myopathy and ALS EMG signals.

myopathy and ALS EMG signals. It is observed that the kurtosis value of ALS EMG signals is higher compared to normal and myopathy EMG signals.

4.6.3   HJORTH PARAMETERS

Hjorth parameter comprises three parameters, namely, activity, complexity and mobility (Hjorth, 1970). These parameters are utilized to describe the statistics on the temporal dynamics of the acquired biosignals (Oh et al., 2014).

The variance of the measured signals $y(t)$ represents the activity parameter and describes the surface of the power spectrum in the frequency domain (Hjorth, 1970).

$$\text{Activity} = \text{var}(y(t)) \qquad (4.7)$$

Complexity parameter compares the similarity of the signal $y(t)$ to a sine wave. If the complexity value is one, then the signals is similar to the sine wave (Hjorth, 1970).

$$\text{Complexity} = \frac{\text{Mobility}\left(\dfrac{dy(t)}{dt}\right)}{\text{Mobility}(y(t))} \qquad (4.8)$$

Mobility parameter describes the proportion of the standard deviation of the power spectrum. (Hjorth, 1970).

$$\text{Mobility} = \sqrt{\frac{\text{var}(y'(t))}{\text{var}(y(t))}} \qquad (4.9)$$

Figure 4.17a shows the activity of normal, myopathy and ALS EMG signals. The activity of the normal EMG signals is higher compared to the myopathy and ALS EMG signals. Further, the activity of normal and abnormal EMG signals does not overlap, but in the case of activity values of abnormal (myopathy and ALS), EMG signals overlap.

The complexity of the ALS EMG signals is higher compared to the complexity of the normal and myopathy EMG signals, as shown in the Figure 4.17b. Further, the variability of the ALS EMG signals is more compared to the normal and myopathy EMG signals. Figure 4.17c shows the Hjorth mobility as a function of normal, myopathy and ALS EMG signals. It is seen that the mobility of the ALS signal is higher compared with the mobility of normal and myopathy electromyograms. Further, the mobility of the abnormal biosignals is higher compared to the normal biosignals. The frequency component is higher for abnormal than normal signals.

4.6.4   HURST EXPONENT

Hurst exponent (H) is a quantification of long-term memory of the time sequence in biological signals and is commonly referred as index of dependence. Hurst exponent

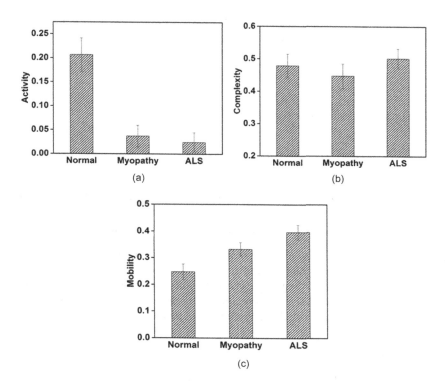

FIGURE 4.17 Hjorth parameters of the normal, myopathy and ALS EMG signals: (a) activity, (b) complexity and (c) mobility.

is a quantification of degree of self-similarity of the time series (Hurst, 1951). If the Hurst exponent value is within the intervals of $0.5 < H < 1$, then the time series can be considered as fractals (Positive autocorrelation). If the Hurst exponent value is equal to 0.5, then the time series follows Brownian motion (uncorrelated series). If the Hurst exponent value is within the interval of $0 < H < 0.5$, then the time series can be considered as negative autocorrelation (Gospodinov et al., 2019). There are two different groups to estimate the Hurst exponent of biological signals, namely, time-based methods and frequency-based methods. Time-based methods include rescaled adjusted range statistics method (R/S method), index of dispersion for counts, variance-time plot and wavelet-based methods. Periodogram method and Whittle method are the frequency-based methods (Gospodinov et al., 2019).

The Hurst exponent is defined as the slope of the straight line that fits $\log(R(m)/S(m))$ as a function of $\log m$ using the least square method (Hurst, 1951).

$$\log\big(R(m)/S(m)\big) = \text{Hurst exponent} * \log m \qquad (4.10)$$

where $S(m)$ is the standard deviation of the partial time series, and the cumulative range of the auxiliary signals $R(m)$ is given as (Hurst, 1951):

$$R(m) = \max_t z[t] - \min_t z[t] \tag{4.11}$$

where $z[t]$ partial cumulative time series is expressed as:

$$z[t] = \sum_{j=0}^{t} \left(x[j] - \bar{x}\right) \tag{4.12}$$

The average of the partial time series with chopped signal length m is given as $\bar{x} = \dfrac{1}{m}\sum_{i=0}^{m-1} x[i]$ (Hurst, 1951). The variation of the Hurst exponent values of normal and abnormal EMG signals, as shown in Figure 4.18. It is found that Hurst exponent of the normal EMG signal is higher compared with the Hurst exponent of the abnormal EMG signals. In all cases, the Hurst exponent value is found is in the range of 0.5–1, indicating long-term positive autocorrelation. Among that normal EMG signals have a higher tendency to cluster to the long range of dependence.

4.6.5 FRACTAL DIMENSION

Fractal dimension (D) is directly correlated to the Hurst exponent and describes the self-similarity or complexity of the signals. The higher value of fractals indicates the higher irregularity of the signals (Falconer, 2004). Several techniques such as Box counting dimension, information dimension, correlation dimension, Higuchi dimension, Lyapunov dimension and Hausdorff dimension are available to estimate the fractal dimension of signals (Falconer, 2004; Boonyakitanont et al., 2020; Hurst, 1951).

The relation between the fractal dimension and Hurst exponent is expressed as:

$$D = 2 - H \tag{4.13}$$

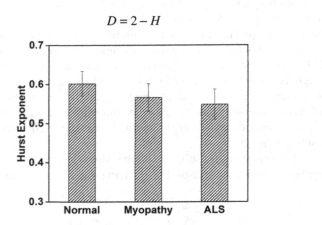

FIGURE 4.18 Hurst exponent of normal, myopathy and ALS EMG signals.

If the Hurst exponent value is zero, then the graph shows a severely curved curve (Boonyakitanont et al., 2020; Hurst, 1951; Blackledge, 2005). If the Hurst exponent value is 1, then the graph tends to be a smooth line. The variation of fractal dimension values of abnormal and normal EMG signals is shown in Figure 4.19. It is seen that the fractal dimension of the ALS signals is higher compared to normal and myopathy EMG signals. The higher fractal value indicates more irregularities. Hence, the fractal dimension of abnormal signals is higher than normal signals.

4.6.6 ZERO CROSSING

Zero crossing (ZC) is a measure of the number of times a signal crosses zero line in the plot, indicating signal frequency approximation (Toledo-Pérez et al., 2020; Waris et al., 2018). The ZC is defined as:

$$ZC = \sum_i f_{ZC}(.) \tag{4.14}$$

The function $f_{ZC}(.)$ can be varied according to the signals. Few methods for estimating the ZC are listed below (Toledo-Pérez et al., 2020; Iqbal & Subramaniam, 2019; Gaudet et al., 2018).

If the two samples are continuous then the ZC can be described as (Toledo-Pérez et al., 2020):

$$f_{ZC}\left(x_i, x_{io+1}\right) = \begin{cases} 1, & x_i > 0 \text{ and } x_{i+1} < 0 \\ & \text{or } x_i < 0 \text{ and } x_{i+1} > 0, \\ 0, & \text{otherwise.} \end{cases} \tag{4.15}$$

If the two samples are consecutive with different signs (when one sample is greater than zero and other sample is less than zero so their product must be negative number), then the ZC function can be described as (Toledo-Pérez et al., 2020):

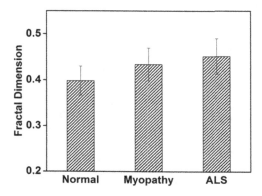

FIGURE 4.19 Fractal dimension of normal, myopathy and ALS EMG signals.

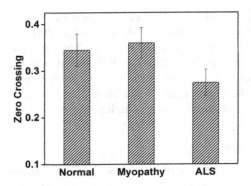

FIGURE 4.20 Variation of ZC values of normal, myopathy and ALS EMG signals.

$$f_{ZC}(x_i, x_{io+1}) = \begin{cases} 1, & x_i \cdot x_{i+1} < 0 \\ 0, & \text{otherwise.} \end{cases} \qquad (4.16)$$

In the above equation, the "sgn" function is included after multiplication to avoid floating point, then the ZC can be rewritten as (Toledo-Pérez et al., 2020):

$$f_{ZC}(x_i, x_{io+1}) = \begin{cases} 1, & \text{sgn}(-x_i \cdot x_{i+1}) > 0 \\ 0, & \text{otherwise.} \end{cases} \qquad (4.17)$$

The variation of ZC values with respect to the normal, myopathy and ALS EMG signals is shown in Figure 4.20. It is observed that the ZC value of myopathy is higher when compared to the ZC values of ALS and normal EMG signals.

4.6.7 SPECTRAL ENTROPY

Figure 4.21 shows the variation of spectral entropy values of normal, myopathy and ALS EMG signals. It is seen that the spectral entropy of myopathy EMG signals is higher compared with the spectral entropy of normal and ALS electromyograms.

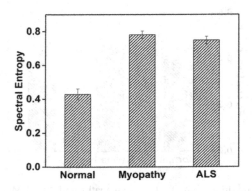

FIGURE 4.21 Variation of spectral entropy of normal, myopathy and ALS EMG signals.

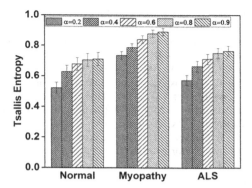

FIGURE 4.22 Variation of Tsallis entropy values of normal, myopathy and ALS EMG signals with respect to different alpha values.

4.6.8 TSALLIS ENTROPY

The variation of Tsallis entropy values of normal and abnormal EMG signals with different α values ranging from 0.2 to 0.9 is shown in Figure 4.22. In all cases, the Tsallis entropy values increase with increasing α values. Further, the Tsallis entropy of myopathy EMG signals is found to be higher compared to the Tsallis entropy of normal and myopathy EMG signals.

Electromyograms are electrical signals acquired from the muscles which are used to diagnose various neuromuscular disorders. The functional phenomenon of the muscles are well described by the measured EMG signals The proper selection of electrode types and the proper localization of electrodes in the muscle regions will improve the accuracy of the overall diagnostic process. Further, the quantitative analysis of EMG signals facilitates the design and development of computer-aided diagnostic systems for the analysis of neuromuscular pathologies.

REFERENCES

Alagumariappan, P., Krishnamurthy, K., Kandiah, S., & Ponnuswamy, M. J. (2017). Effect of electrode contact area on the information content of the recorded electrogastrograms: An analysis based on Rényi entropy and Teager-Kaiser Energy. *Polish Journal of Medical Physics and Engineering, 23*(2), 37–42.

Ambikapathy, B. & Krishnamurthy, K. (2018). Analysis of electromyograms recorded using invasive and noninvasive electrodes: a study based on entropy and Lyapunov exponents estimated using artificial neural networks. *Journal of Ambient Intelligence and Humanized Computing*, 1–9. Doi: 10.1007/s12652-018-0811-6

Ambikapathy, B., Kirshnamurthy, K., & Venkatesan, R. (2018). Assessment of electromyograms using genetic algorithm and artificial neural networks. *Evolutionary Intelligence*, 1–11. Doi: 10.1007/s12065-018-0174-0

Artameeyanant, P., Sultornsanee, S., & Chamnongthai, K. (2016). An EMG-based feature extraction method using a normalized weight vertical visibility algorithm for myopathy and neuropathy detection. *SpringerPlus, 5*(1), 2101.

Artameeyanant, P., Sultornsanee, S., Chamnongthai, K., & Higuchi, K. (2014, December). Classification of electromyogram using vertical visibility algorithm with support vector machine. *In 2014 Asia-Pacific Signal and Information Processing Association Annual Summit and Conference (APSIPA)*, Cambodia (pp. 1–5). IEEE.

Bakiya, A. & Kamalanand, K. (2018, March). Information analysis on electromyograms acquired using monopolar needle, concentric needle and surface electrodes. *In 2018 International Conference on Recent Trends in Electrical, Control and Communication (RTECC)*, Chennai, (pp. 240–244). IEEE.

Bakiya, A., Kamalanand, K., Rajinikanth, V., Nayak, R. S., & Kadry, S. (2020). Deep neural network assisted diagnosis of time-frequency transformed electromyograms. *Multimedia Tools and Applications*, 79(15), 11051–11067.

Begg, R., Lai, D. T., & Palaniswami, M. (2007). *Computational intelligence in biomedical engineering*. Boca Raton, FL: CRC Press.

Blackledge, J. M. (2005). Fractal images and image processing. In: *Digital image processing* (pp. 541–600). Chichester: Horwood Publishing.

Blottner, D. & Salanova, M. (2015). Neuromuscular system. In: *The neuro muscular system: From earth to space life science* (pp. 63–87). Cham: Springer.

Bohr, D. F. (1973). Vascular smooth muscle updated. *Circulation Research*, 32(6), 665–672.

Boon, A. J., Alsharif, K. I., Harper, C. M., & Smith, J. (2008). Ultrasound-guided needle EMG of the diaphragm: technique description and case report. *Muscle and Nerve: Official Journal of the American Association of Electrodiagnostic Medicine*, 38(6), 1623–1626.

Boon, A. J., Oney-Marlow, T. M., Murthy, N. S., Harper, C. M., McNamara, T. R., & Smith, J. (2011). Accuracy of electromyography needle placement in cadavers: Non-guided vs. ultrasound guided. *Muscle and Nerve*, 44(1), 45–49.

Boonyakitanont, P., Lek-Uthai, A., Chomtho, K., & Songsiri, J. (2020). A review of feature extraction and performance evaluation in epileptic seizure detection using EEG. *Biomedical Signal Processing and Control*, 57, 101702.

Couratier, P., Corcia, P., Lautrette, G., Nicol, M., Preux, P. M., & Marin, B. (2016). Epidemiology of amyotrophic lateral sclerosis: a review of literature. *Revue Neurologique*, 172(1), 37–45.

Davis, P. J., Motoyama, E. K., & Cladis, F. P. (2016). Special characteristics of pediatric anesthesia. In: Davis, P. & Cladis, F. (Eds), Smith's Anesthesia for infants and Children E-Book (p. 1376). Maryland, MI: Mosby.

Day, S. (2002). *Important factors in surface EMG measurement* (pp. 1–17). Calgary: Bortec Biomedical Ltd Publishers.

De Luca, C. J. (1997). The use of surface electromyography in biomechanics. *Journal of Applied Biomechanics*, 13(2), 135–163.

Duque, C. J. G., Muñoz, L. D., Mejía, J. G., & Trejos, E. D. (2014, September). Discrete wavelet transform and k-nn classification in EMG signals for diagnosis of neuromuscular disorders. *In 2014 XIX Symposium on Image, Signal Processing and Artificial Vision* (pp. 1–5). IEEE.

Falconer, K. (2004). *Fractal geometry: Mathematical foundations and applications*. Hoboken, NJ: John Wiley & Sons.

Fattah, S. A., Iqbal, M. A., Jumana, M. A., & Doulah, A. S. U. (2012). Identifying the motor neuron disease in EMG signal using time and frequency domain features with comparison. *Signal and Image Processing*, 3(2), 99.

Frontera, W. R. & Ochala, J. (2015). Skeletal muscle: A brief review of structure and function. *Calcified Tissue International*, 96(3), 183–195.

Gandhi, S. M. & Sarkar, B. C. (2016). *Essentials of mineral exploration and evaluation*. Amsterdam: Elsevier.

Garmirian, L. P., Chin, A. B., & Rutkove, S. B. (2009). Discriminating neurogenic from myopathic disease via measurement of muscle anisotropy. *Muscle and Nerve*, 39(1), 16–24.

Gaudet, G., Raison, M., & Achiche, S. (2018). Classification of upper limb phantom movements in transhumeral amputees using electromyographic and kinematic features. *Engineering Applications of Artificial Intelligence*, 68, 153–164.

Geiringer, S. R. (1999). *Anatomic localization for needle electromyography*. Philadelphia, PA: Hanley & Belfus.

Ghapanchizadeh, H., Ahmad, S. A., Ishak, A. J., & Al-quraishi, M. S. (2017). Review of surface electrode placement for recording electromyography signals.

Ghuman, K. & DeBrunner, V. (2014, November). A proof on the invariance of the Hirschman Uncertainty to the Rényi entropy parameter and an observation on its relevance in the image texture classification problem. *In 2014 48th Asilomar Conference on Signals, Systems and Computers*, USA, (pp. 1562–1566). IEEE.

Gospodinov, M., Gospodinova, E., & Georgieva-Tsaneva, G. (2019). Mathematical methods of ECG data analysis. In: Dey, N. & Ashour, A. S. (Eds), *Healthcare data analytics and management* (pp. 177–209). Cambridge, MA: Academic Press.

Greenway, M. J., Andersen, P. M., Russ, C., Ennis, S., Cashman, S., Donaghy, C., ... & Morrison, K. E. (2006). ANG mutations segregate with familial and 'sporadic' amyotrophic lateral sclerosis. *Nature Genetics*, 38(4), 411–413.

Hermens, H. J., Freriks, B., Disselhorst-Klug, C., & Rau, G. (2000). Development of recommendations for SEMG sensors and sensor placement procedures. *Journal of Electromyography and Kinesiology*, 10(5), 361–374.

Hermens, H. J., Freriks, B., Merletti, R., Stegeman, D., Blok, J., Rau, G., ... & Hägg, G. (1999). European recommendations for surface electromyography. *Roessingh Research and Development*, 8(2), 13–54.

Hjorth, B. (1970). EEG analysis based on time domain properties. *Electroencephalography and Clinical Neurophysiology*, 29(3), 306–310.

Hurst, H. E. (1951). Long-term storage capacity of reservoirs. *Transactions of the American Society of Civil Engineers*, 116, 770–799.

Iqbal, N. V. & Subramaniam, K. (2019). Robust feature sets for contraction level invariant control of upper limb myoelectric prosthesis. *Biomedical Signal Processing and Control*, 51, 90–96.

Jost, W. H., Ecker, K. W., & Schimrigk, K. (1994). Surface versus needle electrodes in determination of motor conduction time to the external anal sphincter. *International Journal of Colorectal Disease*, 9(4), 197–199.

Kamen, G. & Gabriel, D. A. (2009). *Essentials of Electromyography*, Human Kinetics Publishers. ISBN: 9781492573593

Leis, A. A. & Trapani, V. C. (2000). *Atlas of electromyography*. Oxford: Oxford University Press.

Lieber, R. L. (2002). *Skeletal muscle structure, function, and plasticity*. Philadelphia, PA: Lippincott Williams & Wilkins.

Lu, H. H. & Jiang, J. (2005). Interface tissue engineering and the formulation of multiple-tissue systems. In: Lee, K. & Kaplan, D. (Eds), *Tissue engineering I* (pp. 91–111). Berlin, Heidelberg: Springer.

MacIntosh, B. R., Gardiner, P. F., & McComas, A. J. (2006). *Skeletal muscle: Form and function*. Champaign, IL: Human Kinetics.

Menkes, D. L. & Pierce, R. (2019). Needle EMG muscle identification: A systematic approach to needle EMG examination. *Clinical Neurophysiology Practice*, 4, 199–211.

Merletti, R. (2000). Surface electromyography: The SENIAM project. *European Journal of Physical and Rehabilitation Medicine*, 36(4), 167.

Merletti, R., Rainoldi, A., & Farina, D. (2001). Surface electromyography for noninvasive characterization of muscle. *Exercise and Sport Sciences Reviews*, 29(1), 20–25.

Nazmi, N., Abdul Rahman, M. A., Yamamoto, S. I., Ahmad, S. A., Zamzuri, H., & Mazlan, S. A. (2016). A review of classification techniques of EMG signals during isotonic and isometric contractions. *Sensors*, 16(8), 1304.

Nikolic, M. (2001). Detailed analysis of clinical electromyography signals: EMG decomposition, findings and firing pattern analysis in controls and patients with myopathy and amytrophic lateral sclerosis, Doctoral dissertation.

Oh, S. H., Lee, Y. R., & Kim, H. N. (2014). A novel EEG feature extraction method using Hjorth parameter. *International Journal of Electronics and Electrical Engineering*, 2(2), 106–110.

Olivetti, G., Cigola, E., Maestri, R., Corradi, D., Lagrasta, C., Gambert, S. R., & Anversa, P. (1996). Aging, cardiac hypertrophy and ischemic cardiomyopathy do not affect the proportion of mononucleated and multinucleated myocytes in the human heart. *Journal of Molecular and Cellular Cardiology, 28*(7), 1463–1477.

Preston, D. C. & Shapiro, B. E. (2012). *Electromyography and neuromuscular disorders e-book: clinical-electrophysiologic correlations (expert consult-online)*. Amsterdam: Elsevier Health Sciences.

Reed, S. M., Bayly, W. M., & Sellon, D. C. (2017). *Equine internal medicine-E-book*. Amsterdam: Elsevier Health Sciences.

Robergs, R. A. & Roberts, S. (1996). *Exercise physiology: Exercise, performance, and clinical applications* (pp. 546–563). St. Louis: Mosby.

Rowland, L. P. & Shneider, N. A. (2001). Amyotrophic lateral sclerosis. *New England Journal of Medicine, 344*(22), 1688–1700.

Sadikoglu, F., Kavalcioglu, C., & Dagman, B. (2017). Electromyogram (EMG) signal detection, classification of EMG signals and diagnosis of neuropathy muscle disease. *Procedia Computer Science, 120*, 422–429.

Saladin, K. S. & McFarland, R. K. (2008). The circulatory system II: Heart. In: *Human Anatomy* (4th ed., pp. 572–594) New York: McGraw-Hill.

Singh, V. P. (2013). *Entropy theory and its application in environmental and water engineering*. Hoboken, NJ: John Wiley & Sons.

Singh, V. P. (2014). *Entropy theory in hydrologic science and engineering*. New York: McGraw Hill Professional.

Smelser, N. J. & Baltes, P. B. (Eds.). (2001). *International encyclopedia of the social & behavioral sciences* (Vol. 11). Amsterdam: Elsevier.

Smith, L. H. & Hargrove, L. J. (2013, July). Comparison of surface and intramuscular EMG pattern recognition for simultaneous wrist/hand motion classification. In *2013 35th Annual International Conference of the IEEE Engineering in Medicine and Biology Society (EMBC)*, Japan, (pp. 4223–4226). IEEE.

Soderberg, G. L. & Cook, T. M. (1984). Electromyography in biomechanics. *Physical Therapy, 64*(12), 1813–1820.

Tenan, M. S., Tweedell, A. J., & Haynes, C. A. (2017). Analysis of statistical and standard algorithms for detecting muscle onset with surface electromyography. *PLoS One, 12*(5), e0177312.

Thorpe, C. T., Birch, H. L., Clegg, P. D., & Screen, H. R. (2013). The role of the non-collagenous matrix in tendon function. *International Journal of Experimental Pathology, 94*(4), 248–259.

Toledo-Pérez, D. C., Rodríguez-Reséndiz, J., & Gómez-Loenzo, R. A. (2020). A study of computing zero crossing methods and an improved proposal for EMG signals. *IEEE Access, 8*, 8783–8790.

Van Boxtel, A. (2001). Optimal signal bandwidth for the recording of surface EMG activity of facial, jaw, oral, and neck muscles. *Psychophysiology, 38*(1), 22–34.

Wang, W. Z., Fang, X. H., Stephenson, L. L., Khiabani, K. T., & Zamboni, W. A. (2008). Ischemia/reperfusion-induced necrosis and apoptosis in the cells isolated from rat skeletal muscle. *Journal of Orthopaedic Research, 26*(3), 351–356.

Waris, A., Niazi, I. K., Jamil, M., Gilani, O., Englehart, K., Jensen, W., ... & Kamavuako, E. N. (2018). The effect of time on EMG classification of hand motions in able-bodied and transradial amputees. *Journal of Electromyography and Kinesiology, 40*, 72–80.

Zawawi, T. T., Abdullah, A. R., Shair, E. F., Halim, I., & Rawaida, O. (2013, December). Electromyography signal analysis using spectrogram. In *2013 IEEE Student Conference on Research and Development (SCOReD)*, Malaysia, (pp. 319–324). IEEE.

5 Activity of Muscles in Non-Myopathic Conditions
A Case Study from Lymphatic Filariasis

5.1 FILARIASIS AND LYMPHOEDEMA

Filariasis is well documented since the mid-19th century and is caused by thread-like parasitic roundworms *Wuchereria bancrofti* and *Brugia malayi* (Otsuji, 2011, Babu & Nutman, 2012; Nutman et al., 2013). The early stage of filariasis symptoms are recurrent fever episodes with chills, inflammation of the lymph nodes and erythematic lesions on the skin. Though the infestation occurs during childhood, most remain asymptomatic (Pani et al., 1991; Witt & Ottesen, 2001; Freedman et al., 1994). It is established that adult worms cause lymphatic damage, leading to lymphatic obstruction and stasis (Taylor, 2003). Lymphatic obstruction, in turn, predisposes to recurrent bacterial infection of the skin and soft tissues of the draining area. Lymphatic vessels of the scrotum and lower extremities are commonly involved in filarial infection. Recurrent infections followed by healing and lymph stasis are considered as the major factors for the progression of the lymphoedema of the extremities (Walther & Muller, 2003; Shenoy, 2002; Shenoy et al., 2007). Lymphoedema affecting the foot causes intertrigo of the toe-web spaces and several predisposing factors are attributed for the development of the intertrigo (De Britto et al., 2015). For morbidity management and disability prevention in National filariasis control program of disease-endemic countries, the clinical staging of the lymphoedema of the extremity is done based on the reversibility of the lymphoedema on overnight rest, skin thickening (due to fibrosis) and the secondary skin changes such as nodules, abscess formation, skin ulcers, warts and intertrigo. In brief, in stage-1, the lymphoedema is completely reversible on overnight rest, in stage-2, it is partially reversible, in stage-3, skin thickening is perceptible, and in stage-4, and multiple skin lesions are visible (Figure 5.1).

It is expected that the fibrotic changes in lymphedema interfere with the output in EMG signals, and the inferences derived from these changes may serve as a surrogate marker to assess the fibrotic changes in different grades on lymphedema, as well as to assess the effect of conservative and surgical management.

Electromyography is the electrodiagnostic technique for recording and analyzing the electrical potentials generated from the neuromuscular system

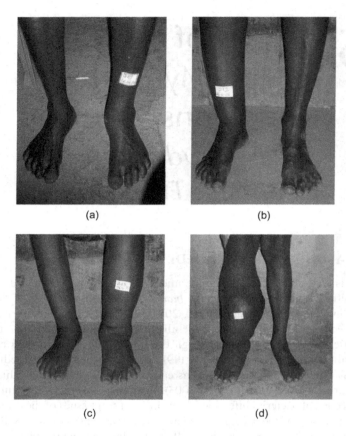

(a)   (b)

(c)   (d)

FIGURE 5.1   Stages of lymphoedema (a) lymphoedema – stage-1 (left leg), (b) lymphoedema – stage-2 (right leg), (c) lymphoedema – stage-3 (left leg), (d) lymphoedema – stage-4 (right leg).

(Ambikapathy & Krishnamurthy, 2018; Ambikapathy et al., 2018; Bakiya & Kamalanand, 2018; Bakiya et al., 2020). The recorded electrical potentials are utilized for the diagnosis of several neuromuscular disorders such as myopathy, multiple sclerosis, peripheral neuropathy and amyotrophic lateral sclerosis (Sadikoglu et al., 2017; Duque et al., 2014), as discussed in the earlier chapters. In general, the EMG recording instruments are designed to record electrical signals in the frequency range of 20 and 450 Hz. However, the frequency range of healthy EMG signals is between 30 and 60 Hz (Duque et al., 2014; Subasi, 2012).

Feature extraction is the process of extracting useful information from acquired EMG signals, and such extracted features of EMG signals define the characteristics of the original EMG signals (Acharya et al., 2011). The feature extraction techniques for EMG signals are broadly classified into three categories, namely, the time-domain, frequency-domain and time-frequency techniques (Nazmi et al., 2016; Phinyomark et al., 2011, 2012). In this study, the EMG signals acquired from normal and fibrotic lymphoedema patients are analyzed using three different prominent feature extraction techniques such as time-domain, frequency-domain and time-frequency features.

5.2 ACQUISITION OF EMG SIGNALS

Normal and fibrotic EMG signals were acquired using an EMG acquisition device over 30 s. A band-pass filter with a pass band of 20–500 Hz was used to filter the noise. The sampling rate of 1,000 Hz was considered for the measurement of EMG signals from normal subjects and patients with fibrotic lymphoedema. The signals were acquired from the subjects in sitting position. Figure 5.2 shows the typical EMG signals acquired from the normal leg and the fibrotic lymphedema leg of the patients from an endemic city. It is observed that the pattern of the EMG signals of the fibrotic lymphedema leg is different compared to the EMG signals from the normal leg of the same individual. The peak amplitude of the fibrotic EMG signals was observed to be 0.02 V, whereas, in the case of normal EMG signals, the peak amplitude was 0.2 V.

5.3 TIME-DOMAIN ANALYSIS OF NORMAL AND FIBROTIC EMG SIGNALS

The acquired EMG signals were analyzed using time-domain features such as the Hjorth parameters (Hjorth activity, Hjorth complexity and Hjorth mobility), Hurst exponent, fractal dimension, entropy, descriptive statistics such as mean median

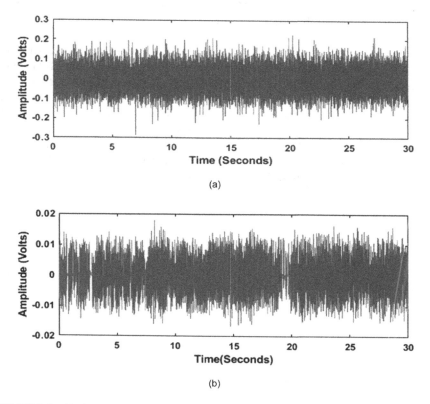

(a)

(b)

FIGURE 5.2   Typical EMG signals acquired from (a) normal leg and (b) fibrotic leg.

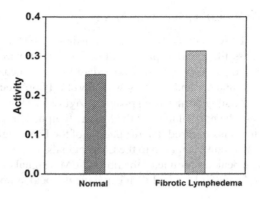

FIGURE 5.3   Activity of EMG signals acquired from normal and fibrotic leg.

and skewness and zero crossing. Furthermore, features such as the myopulse percentage rate (Shair et al., 2017; AlOmari & Liu, 2014) and Willison amplitude (Phinyomark et al., 2008; Negi et al., 2016) were extracted from EMG signals acquired from normal and fibrotic regions of the leg and were analyzed. Figure 5.3 shows the Hjorth activity of the EMG signals recorded from normal and fibrotic regions. It is observed that the mean activity of the fibrotic EMG signals is higher compared to the mean activity of the normal EMG signals. The variation of the complexity of the EMG signals of normal and fibrotic regions is shown in Figure 5.4. It is seen that the value of complexity is higher in normal EMG signals compared to the complexity of the EMG signals acquired from fibrotic regions. Figure 5.5 shows the variation of mean mobility values of the normal EMG signals and the signals acquired from fibrotic regions of the leg. It is observed that the mean mobility value is higher in EMG signals acquired from the fibrotic leg compared to the mean mobility of the EMG signals acquired from the normal leg. Further, the frequency component is higher in the case of EMG signals recorded from fibrotic leg compared to the normal EMG signals.

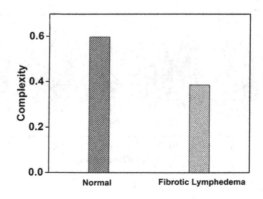

FIGURE 5.4   Complexity of EMG signals acquired from normal and fibrotic leg.

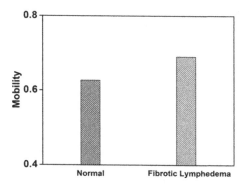

FIGURE 5.5   Mobility of EMG signals acquired from normal and fibrotic leg.

The variation of Hurst exponent of the normal and fibrotic EMG signals is shown in Figure 5.6. The Hurst exponent value of the EMG signals of the normal legs is higher compared to that of the EMG signals acquired from the fibrotic regions of the lymphedema leg. The EMG signals in normal legs exhibit positive autocorrelation (Hurst exponent value is between 0.5 and 1), and fibrotic lymphedema EMG signals exhibit negative autocorrelation. (Hurst exponent value is <0.5). Figure 5.7 shows the variation of the fractal dimension of normal and fibrotic EMG signals. The fractal dimension is higher in the case of fibrotic leg EMG signals when compared to the normal EMG signals. The variation of the entropy of normal and fibrotic EMG signals is shown in Figure 5.8. The mean entropy values of normal EMG signals are higher compared to the mean entropy values of fibrotic EMG signals.

Figures 5.9 and 5.10 show the variation of mean and median of normal and fibrotic EMG signals, respectively. It is found that the mean and median values are higher in the case of EMG signals of normal leg compared to the EMG signals of the fibrotic regions of lymphoedema leg. Figure 5.11 shows the variation of skewness of normal EMG and EMG recorded at the fibrotic regions of the lymphoedema leg. It is seen that the skewness of the fibrotic EMG signals is higher compared to the skewness of the normal EMG signals. The skewness value of the normal and fibrotic EMG signals are negatively skewed.

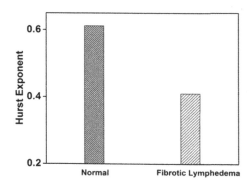

FIGURE 5.6   Hurst exponent of EMG signals acquired from normal and fibrotic leg.

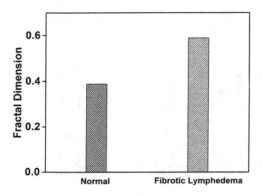

FIGURE 5.7   Fractal dimension of EMG signals acquired from normal and fibrotic leg.

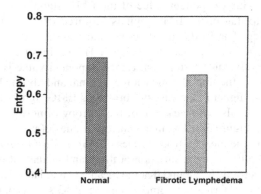

FIGURE 5.8   Entropy of EMG signals acquired from normal and fibrotic leg.

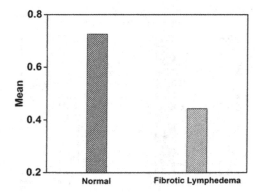

FIGURE 5.9   Mean of EMG signals acquired from normal and fibrotic leg.

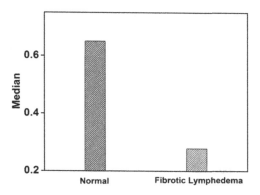

FIGURE 5.10 Median of EMG signals acquired from normal and fibrotic leg.

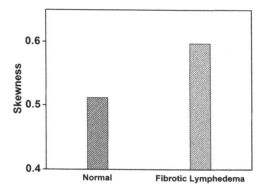

FIGURE 5.11 Skewness of EMG signals acquired from normal and fibrotic leg.

The variation of zero crossing value of normal and fibrotic EMG signals is shown in Figure 5.12. The zero crossing of the EMG signals acquired from fibrotic leg is higher compared to the zero crossing of the normal EMG signals.

Figure 5.13 shows the myopulse percentage rate of normal and fibrotic EMG signals. The myopulse percentage rate of normal EMG signals is higher compared to the myopulse percentage rate of EMG signals recorded from fibrotic regions. The variation of Willison amplitude of the normal and fibrotic EMG signals is shown in Figure 5.14. The Willison amplitude value is higher in the case of normal EMG signals compared to the Willison amplitude of fibrotic EMG signals.

5.4 FREQUENCY-DOMAIN ANALYSIS OF NORMAL AND FIBROTIC ELECTROMYOGRAMS

The frequency-domain analysis of the normal EMG signals and the EMG signals acquired from patients with fibrotic changes was performed using the fast Fourier transform (FFT) and spectral entropy. Figure 5.15 shows the FFT of the EMG signals acquired from normal and fibrotic legs. The frequency components of EMG signals

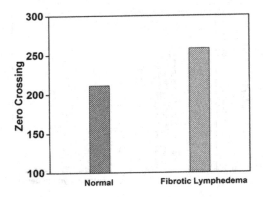

FIGURE 5.12   Zero crossing of EMG signals acquired from normal and fibrotic leg.

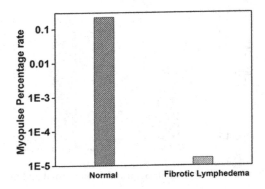

FIGURE 5.13   Myopulse percentage rate of EMG signals acquired from normal and fibrotic leg.

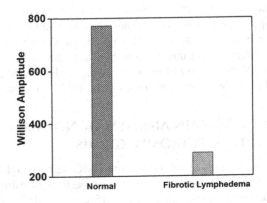

FIGURE 5.14   Willison amplitude of EMG signals acquired from normal and fibrotic leg.

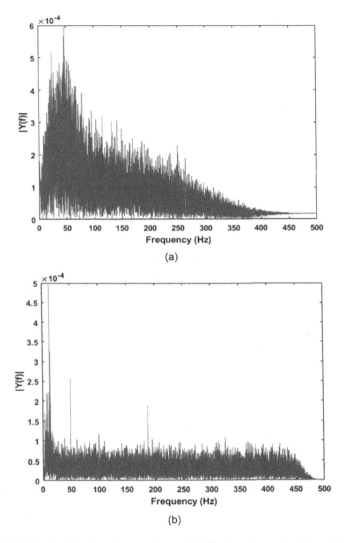

FIGURE 5.15 Typical fast Fourier transform of EMG signals acquired from (a) normal leg and (b) fibrotic leg.

are in the range of 0 and 500 Hz. The peak frequency of the normal EMG signals is 47.88 Hz and fibrotic EMG signals is 13.24 Hz. Further, the variation of spectral entropy values of the normal and fibrotic EMG signals is shown in Figure 5.16. It is seen that the spectral entropy value of EMG signals acquired from fibrotic regions is higher compared to the spectral entropy value of normal EMG signals.

Electromyogram is most commonly applied to diagnose and evaluate familial myopathies and to assess the impact of systemic diseases such as endocrine disorders. Occasionally, it is also applied in autoimmune inflammatory diseases such as

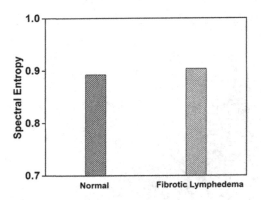

FIGURE 5.16 Spectral entropy of EMG signals acquired from normal and fibrotic leg.

rheumatoid arthritis. However, it is rarely applied to assess the impact on localized conditions and to draw valuable conclusions from those results. For example, the impact of the activator treatment on orbicularis oris and mentalis muscles was assessed by De Souza et.al. (2008) who assessed the EMG changes of the lower lip following the incremental lower lip advancement on intraoral pressure and electromyographic (EMG) activity of the lower lip (Jack et al., 2014). In lymphatic filariasis, stage-3 and 4 patients have extensive fibrosis of the skin and soft tissues and EMG with surface electrodes on appropriate sites also may throw light on the fibrotic changes.

REFERENCES

Acharya, U. R., Ng, E. Y. K., Swapna, G., & Michelle, Y. S. L. (2011). Classification of normal, neuropathic, and myopathic electromyograph signals using nonlinear dynamics method. *Journal of Medical Imaging and Health Informatics, 1*(4), 375–380.

Alomari, F. & Liu, G. (2014). Analysis of extracted forearm sEMG signal using LDA, QDA, K-NN classification algorithms. *The Open Automation and Control Systems Journal, 6*(1), 108–116.

Ambikapathy, B. & Krishnamurthy, K. (2018). Analysis of electromyograms recorded using invasive and noninvasive electrodes: A study based on entropy and Lyapunov exponents estimated using artificial neural networks. *Journal of Ambient Intelligence and Humanized Computing*, 1–9. Doi: 10.1007/s12652-018-0811-6

Ambikapathy, B., Kirshnamurthy, K., & Venkatesan, R. (2018). Assessment of electromyograms using genetic algorithm and artificial neural networks. *Evolutionary Intelligence*, 1–11. Doi: 10.1007/s12065-018-0174-0

Babu, S. & Nutman, T. B. (2012, November). Immunopathogenesis of lymphatic filarial disease. In: Miguel Stadecker (Eds.) *Seminars in immunopathology* (Vol. 34, No. 6, pp. 847–861). Berlin, Germany: Springer-Verlag.

Bakiya, A. & Kamalanand, K. (2018, March). Information analysis on electromyograms acquired using monopolar needle, concentric needle and surface electrodes. *In 2018 International Conference on Recent Trends in Electrical, Control and Communication (RTECC)*, Chennai, India (pp. 240–244). IEEE.

Bakiya, A., Kamalanand, K., Rajinikanth, V., Nayak, R. S., & Kadry, S. (2020). Deep neural network assisted diagnosis of time-frequency transformed electromyograms. *Multimedia Tools and Applications, 79*(15), 11051–11067.

De Britto, L. J., Yuvaraj, J., Kamaraj, P., Poopathy, S., & Vijayalakshmi, G. (2015). Risk factors for chronic intertrigo of the leg in Southern India: A case-control study. *The International Journal of Lower Extremity Wounds, 14*(4), 377–383. Doi: 10.1177/1534734615604289.

De Souza, D. R., Semeghini, T. A., Kroll, L. B., & Berzin, F. (2008). Oral myofunctional and electromyographic evaluation of the orbicularis oris and mentalis muscles in patients with class II/1 malocclusion submitted to first premolar extraction. *Journal of Applied Oral Science: Revista FOB, 16*(3), 226–231. Doi: 10.1590/s1678-77572008000300012.

Duque, C. J. G., Muñoz, L. D., Mejía, J. G., & Trejos, E. D. (2014, September). Discrete wavelet transform and k-nn classification in EMG signals for diagnosis of neuromuscular disorders. *In 2014 XIX Symposium on Image, Signal Processing and Artificial Vision (STSIVA)*, Colombia (pp. 1–5). IEEE.

Freedman, D. O., de Alemeida Filho, P. J., Besh, S., Maia e Silva, M. C., Braga, C., Maciel, A. (1994). Lymphoscintigraphic analysis of lymphatic abnormalities in symptomatic and asymptomatic human filariasis. *Journal of Infectious Diseases, 170*, 927–933.

Jack, H. C., Kieser, J., Antoun, J. S., & Farella, M. (2014). The effect of incremental lower lip advancement on oral pressure and EMG activity of the lower lip. *European Journal of Orthodontics, 36*(6), 672–677. Doi: 10.1093/ejo/cjt094.

Nazmi, N., Abdul Rahman, M., Yamamoto, S. I., Ahmad, S., Zamzuri, H., & Mazlan, S. (2016). A review of classification techniques of EMG signals during isotonic and isometric contractions. *Sensors, 16*(8), 1304.

Negi, S., Kumar, Y., & Mishra, V. M. (2016, September). Feature extraction and classification for EMG signals using linear discriminant analysis. *In 2016 2nd International Conference on Advances in Computing, Communication, & Automation (ICACCA) (Fall)*, Bareilly, India (pp. 1–6). IEEE.

Nutman, T. B. (2013). Insights into the pathogenesis of disease in human lymphatic filariasis. *Lymphatic Research and Biology, 11*(3), 144–148.

Otsuji, Y. (2011). History, epidemiology and control of filariasis. *Tropical Medicine and Health, 39*(1 Suppl 2), 3.

Pani, S. P., Balakrishnan, N., Srividya, A., Bundy, D. A. P., & Grenfell, B. T. (1991). Clinical epidemiology of bancroftian filariasis: Effect of age and gender. *Transactions of the Royal Society of Tropical Medicine and Hygiene, 85*, 260–264.

Phinyomark, A., Hirunviriya, S., Nuidod, A., Phukpattaranont, P., & Limsakul, C. (2011). Evaluation of EMG feature extraction for movement control of upper limb prostheses based on class separation index. *In 5th Kuala Lumpur International Conference on Biomedical Engineering* (pp. 750–754). Springer, Berlin, Heidelberg.

Phinyomark, A., Limsakul, C., & Phukpattaranont, P. (2008, December). EMG feature extraction for tolerance of white Gaussian noise. *In Proceedings of International Workshop and Symposium Science Technology*, Thailand (pp. 178–183).

Phinyomark, A., Phukpattaranont, P., & Limsakul, C. (2012). Feature reduction and selection for EMG signal classification. *Expert Systems with Applications, 39*(8), 7420–7431.

Sadikoglu, F., Kavalcioglu, C., & Dagman, B. (2017). Electromyogram (EMG) signal detection, classification of EMG signals and diagnosis of neuropathy muscle disease. *Procedia Computer Science, 120*, 422–429.

Shair, E. F., Ahmad, S. A., Marhaban, M. H., Mohd Tamrin, S. B., & Abdullah, A. R. (2017). EMG processing based measures of fatigue assessment during manual lifting. *BioMed Research International*, 2017, 3937254.

Shenoy, R. K. (2002). Management of disability in lymphatic filariasis--an update. *The Journal of Communicable Diseases, 34*(1), 1–14.

Shenoy, R. K., Suma, T. K., Kumaraswami, V., Padma, S., Rahmah, N., Abhilash, G., & Ramesh, C. (2007). Doppler ultrasonography detects adult worm nests in lymph vessels of children with brugian filariasis. *Annals of Tropical Medicine and Parasitology, 101*, 173–180.

Subasi, A. (2012). Classification of EMG signals using combined features and soft computing techniques. *Applied Soft Computing, 12*(8), 2188–2198.

Taylor, M. J. (2003). Wolbachia in the inflammatory pathogenesis of human filariasis. *Annals of the New York Academy of Sciences, 990*(1), 444–449.

Walther, M. & Muller, R. (2003). Diagnosis of human filariases (except onchocerciasis).

Witt, C. & Ottesen, E. A. (2001). Lymphatic filariasis: An infection of childhood. *Tropical Medicine and International Health, 6*, 582–606. Doi: 10.1046/j.1365-3156.2001.00765.x.

Index

Printed in the United States
by Baker & Taylor Publisher Services